Anguel S. Stefanov and Marco Giovanelli
Editors

D0888578

General Relativity
1916 - 2016

Selected peer-reviewed papers presented at the Fourth
International Conference on the Nature and Ontology
of Spacetime, dedicated to the 100th anniversary of the
publication of General Relativity, 30 May - 2 June 2016,
Golden Sands, Varna, Bulgaria

MINKOWSKI
Institute Press

Anguel S. Stefanov
Institute for the Study of Societies and Knowledge
Bulgarian Academy of Sciences
13A Moskovska Street
1000 Sofia, Bulgaria

Marco Giovanelli
Universität Tübingen, Forum Scientiarum
Doblerstrae 33
72074 Tübingen, Germany

Cover: Albert Einstein at the time of his 1921 Nobel Prize; Einstein received it one year later in 1922 (`https://www.nobelprize.org/nobel_prizes/physics/laureates/1921/einstein-facts.html`)

© Minkowski Institute Press 2017

ISBN: 978-1-927763-46-9 (softcover)
ISBN: 978-1-927763-47-6 (ebook)

Minkowski Institute Press
Montreal, Quebec, Canada
http://minkowskiinstitute.org/mip/

For information on all Minkowski Institute Press publications visit our website at http://minkowskiinstitute.org/mip/books/

Preface

The present volume is composed of selected peer-reviewed papers presented at the *Fourth International Conference on the Nature and Ontology of Spacetime* dedicated to the 100th anniversary of the publication of General Relativity, held in Golden Sands near Varna, Bulgaria, 30 May – 2 June 2016. The first three volumes contained contributions by participants in the first three spacetime conferences, which were held in Montreal, Canada.

The papers included in the first part of the volume are dedicated mainly to conceptual aspects of spacetime, some of which concerning ontological and historical issues about the genesis and the understanding of the concept "spacetime." Based on the alleged positive result of the Miller ether-drift experiment and the history of similar experiments with negative results, M. Giovanelli analyzes the difference between the so called kinematical "Einstein contraction" and dynamical "Lorentz contraction" of the length of moving bodies, proposed by Reichenbach. V. Petkov's contribution contains interesting conceptual and historical arguments that, had he lived longer, Minkowski could have referred the cause of gravitation to the non-Euclidean geometry of spacetime before Einstein. Two of the papers, those by A. Foligno and A. Stefanov, examine a philosophical setting concerning a basic ontological quality of space and spacetime – its relational or substantival nature. The curious theme about the possibility of time travel is not forgotten, but competently elucidated by R. A. Mulder and D. Dieks through the explication that a specific kind of determinism arises along with the strange features of closed time-like curves. A clear solution to the difficult problem about the meaning of spacetime singularities is presented in the paper by O. C. Stoica.

Papers considering a quantum based source of gravitation as well as physical and informational aspects of spacetime comprise the second part of the volume. The basic claim in D. Dieks' paper is that, in spite of some other interpretations, no one of the two theories AdS/CFT (anti-de Sitter quantum theory of gravity and conformal field theory), which exhibit an exact duality is more fundamental than the other one; and this implies that gravity does not emerge from a more fundamental layer of reality that is gravitationless and non-spatial. Tools from quantum information/quantum computation are used in the paper of K. Irwin, M. M. Amaral, R. Aschheim, F. Fang for the sake of better understanding of the spin networks in Loop Quantum Gravity. Thus a new project is outlined for application of tools like Discrete Quantum Walk to spacetime. The ontological status of the Unruh effect is the subject matter of A. Dobado's contribution. A concrete calculation of the effects of the Unruh effect within a specific model is presented and the conclusion is reached that the Unruh effect is an ontic effect, rather than an epistemic one. The recent popularity of the Everett's interpretation of quantum mechanics was taken into account by L. Marchildon for the sake of reasoning how the multiplicity of emerging universes reflects upon the different views of spacetime.

Formal approaches to spacetime are presented in the third part. No background spacetime is assumed within T. Benda's mathematical scheme, but as an initial step worldliness are taken to be primitive entities. Then a conservative extension of Zermelo's set theory based on the relation of intersection between worldlines is used for the reconstruction of fundamental properties of spacetime. Petr Švarný's contribution is dedicated to a modern approach of attracting temporal logics for depicting time order. The idea of a "Copernican Turn" in the paper is the introduction of an observer in three realizations of logical models, while showing that this operation allows for the same evaluation of statements in both temporal and atemporal logics.

Federico Silvagni's paper "Spacetime in language" is situated in a separate part, because of its intention to demonstrate possible implications of the conception of events as spacetime points for solving specific questions in linguistics.

We hope that the contributions presented within this volume, which concern the nature and ontology of spacetime, demonstrate both the theoretical importance of the concept of spacetime for the developing scientific knowledge, and the need for future investigations on this topic.

25 April 2017   The editors

CONTENTS

II SPACETIME AND QUANTUM THEORY   101

7 Duality and Emergence
Dennis Dieks

8 Quantum Walk on Spin Network and the Golden Ratio as the Fundamental Constant of Nature
Klee Irwin, Marcelo M. Amaral, Raymond Aschheim, Fang Fang

9 Spontaneous symmetry breaking and the Unruh effect
Antonio Dobado

Contributors

Marcelo M. Amaral
Quantum Gravity Research
Los Angeles, CA
and
Institute for Gravitation and the Cosmos,
Pennsylvania State University, University Park, PA
mramaciel@gmail.com

Raymond Aschheim
Quantum Gravity Research
Los Angeles, CA
Raymond@QuantumGravityResearch.org

Thomas Benda
Institute of Philosophy of Mind
National Yang Ming University
Taipei, Taiwan
tbenda@ym.edu.tw

Dennis Dieks
History and Philosophy of Science
Utrecht University
Utrecht, The Netherlands

Antonio Dobado
Departamento de Física Teórica I
Universidad Complutense, 28040 Madrid, Spain

Fang Fang
Quantum Gravity Research
Los Angeles, CA
Fang@QuantumGravityResearch.org

Antonella Foligno
University of Urbino
antonella.foligno@uniurb.it

Marco Giovanelli
Universität Tübingen, Forum Scientiarum
Doblerstrae 33
72074 Tübingen, Germany

Klee Irwin
Quantum Gravity Research
Los Angeles, CA
Klee@QuantumGravityResearch.org

Louis Marchildon
Département de chimie, biochimie et physique,
Université du Québec, Trois-Rivières, Qc. Canada G9A 5H7
louis.marchildon@uqtr.ca

Ruward A. Mulder
History and Philosophy of Science
Utrecht University
Utrecht, The Netherlands

Vesselin Petkov
Minkowski Institute
Montreal, Quebec, Canada
http://minkowskiinstitute.org/
vpetkov@minkowskiinstitute.org

Federico Silvagni
Universidad Complutense de Madrid
Madrid, Spain
f.silvagni@ucm.es

Anguel S. Stefanov
Institute for the Study of Societies and Knowledge
Bulgarian Academy of Sciences
13A Moskovska Street
1000 Sofia, Bulgaria

Ovidiu Cristinel Stoica
Department of Theoretical Physics
National Institute of Physics and Nuclear Engineering – Horia Hulubei
Bucharest, Romania
cristi.stoica@theory.nipne.ro, holotronix@gmail.com

Petr Švarný
Charles University
Prague

Part I

SPACETIME: PHILOSOPHICAL, CONCEPTUAL, AND HISTORICAL ASPECTS

A. S. Stefanov, M. Giovanelli (Eds), *General Relativity 1916 - 2016. Selected peer-reviewed papers presented at the Fourth International Conference on the Nature and Ontology of Spacetime, dedicated to the 100th anniversary of the publication of General Relativity, 30 May - 2 June 2016, Golden Sands, Varna, Bulgaria* (Minkowski Institute Press, Montreal 2017). ISBN 978-1-927763-46-9 (softcover), ISBN 978-1-927763-47-6 (ebook).

1   Lorentz Contraction vs. Einstein Contraction. Reichenbach and the Philosophical Reception of Miller's Ether-Drift Experiments

Marco Giovanelli

Abstract   In 1925 the American experimentalist Dayton C. Miller [35] published the results of a series of Michelson-Morely-type of experiments that he conducted at the top of Mount Wilson, in Southern California. Miller's "scandalous" detection of an ether-drift sparked, as one would expect, considerable debate, which went well beyond the physics community, about a possible refutation of special relativity. An account of the role of Miller's experiments in the history of ether-drift experiments has been provided in the classical monograph of Loyd S. Swenson [72]. Einstein's attitude towards Miller's result has been investigated by Klaus Hentschel [21]. The debate that ensued in the physics community, in particular in the American one, has been recently reconstructed by Roberto Lalli [30]. However, the reception of Miller's ether-drift detection among "philosophers" is still to be written. This paper attempts to make a first step towards closing this gap in the literature.

1   Introduction

Reichenbach was the first to join the debate. Still in 1925, he published a paper [48] in which he offered a first philosophical evaluation of Miller's findings with the intention to demonstrate the physical implications of his axiomatization of special relativity [47]. The peculiarity of Reichenbach's approach can be inferred from some remarks that Moritz Schlick sent to Einstein at the end of 1925 (Schlick to Einstein, Dec. 27, 1925; [9, 21-591]; cf. [20, 361f.]). To his surprise, Schlick found conclusions in Reichenbach's paper that could not have been further from his own. Reichenbach did not believe that the Lorentz contraction was an *ad hoc* hypothesis, which Einstein had dispelled [see 26, on this issue]. On the contrary, he believed that the contraction needed a

A. S. Stefanov, M. Giovanelli (Eds), *General Relativity 1916 - 2016. Selected peer-reviewed papers presented at the Fourth International Conference on the Nature and Ontology of Spacetime, dedicated to the 100th anniversary of the publication of General Relativity, 30 May - 2 June 2016, Golden Sands, Varna, Bulgaria* (Minkowski Institute Press, Montreal 2017). ISBN 978-1-927763-46-9 (softcover), ISBN 978-1-927763-47-6 (ebook).

dynamical explanation based on a theory of matter in special relativity, just like in Lorentz's theory. Thus Schlick realized that for Reichenbach there were no differences between Lorentz and Einstein's theories, and therefore no conventional choice between them. Moreover, the acceptance of Miller's results would not have implied a return to the old ether theory. By reading Schlick's letter, one is immediately disabused of the conviction that Reichenbach's axiomatization was a mathematically sophisticated version of Schlick's conventionalist reading of special relativity.

Reichenbach's line of argument was based on the distinction between to type of contractions, the "Lorentz contraction" and the "Einstein contraction", a distinction which would resurface in his major 1928 monograph on the philosophy of space and time [52]. In a slightly updated parlance, one can say that the Lorentz contraction compares the proper length of a moving rod in special relativity with the length that the rod would have had in the classical theory; the Einstein contraction compares the proper length of a relativistic rod with its coordinate length in a moving frame. The first contraction implies a real difference and requires a molecular-atomistic explanation in both Einstein and Lorentz's theories. The second contraction is a perspectival difference that depends on the definition of simultaneity. Reichenbach held the controversial opinion that only the first contraction is necessary to explain the Michelson experiment. The present paper intends to show that Reichenbach's distinction between two types of rod contractions, even if it is seldom mentioned in the literature, is possibly more characteristic of Reichenbach's interpretation of special relativity than than his thesis of the conventionality of simultaneity—the freedom to choose which events are simultaneous in a given inertial frame depending on the value of the parameter ϵ—which, on the contrary, has been subject of a long-standing and sometimes heated debate [cf. 24, sec. 4 for an overview].

Section 2 of this paper presents a sketch of Reichenbach's axiomatization of special relativity, which is indispensable to understanding his line of argument. Section 3 introduces the distinction between Lorentz and Einstein contractions. Section 4 shows how Reichenbach uses the distinction to provide a philosophical account of Miller's experiment and section 5 analyzes Schlick's reaction to Reichenbach's approach. Section 6 follows Reichenbach's "geometrization" of the opposition between Lorentz/Einstein-contraction in terms of Minkowski diagrams, as presented in his 1928 book. The paper concluded by showing how Reichenbach's distinction was resurrected in the 1950s by Adolf Grünbaum [17], who used it to "defend" the Lorentz contraction from Karl Popper's ad-hocness charge [18].

In this sense, it was Reichenbach the first who attempted to move the debate over the relationship between Einstein and Lorentz's theories out of the arena of the *ad hoc*/non-*ad hoc* distinction [26], and to frame it in the now familiar terms of the "arrow of explanation". According to Reichenbach, both theories attempt to account for an odd coincidence, that matter and fields "contract" in the same way (that is in Reichenbach's parlance the light geometry and matter geometry agree). Lorentz explained this odd coincidence as a deviation of rods and clocks from their natural, non-relativistic behavior. Einstein, on the contrary, refused to give an explanation and simply declared that the relativistic behavior of rods and clocks to be the natural one. According to Reichenbach, however, an explanation is needed in Einstein theory, just like in Lorentz theory. Surprisingly, in Reichenbach's view the explanation is not to be searched in the geometrical structure of spacetime [28], but in a theory of the material structure of rods and clocks. In this sense, Reichenbach's approach can be regarded as a curious anticipation of what we would call a neo-Lorentzian approach to special relativity Brown [1].

2 Reichenbach's Axiomatization and the Role of the Michelson Experiment

On September 17, 1921 Reichenbach wrote to Moritz Schlick that in a few days he was going to present his project of an axiomatization of special relativity in Jena, at the meeting of the *Gesellschaft Deutscher Naturforscher und Ärzte* (Reichenbach to Schlick, Sep. 17, 1921; [64]). As he went on to explain to Einstein immediately thereafter, the main ambition of the project was to show that in special relativity "one can get along without rigid rods and material clocks" by using only light rays (Einstein to Reichenbach, Oct. 12, 1921; [14, Vol. 12, Doc. 266]). Reichenbach's first sketch of his axiomatization project was published in the same year [45]. Reichenbach sent it to on at the beginning of 1922, warning that, although the paper was inspired by conventionalism, its results "reveals those facts that also conventionalism cannot interpret" (Reichenbach to Schlick, Jan. 18, 1922; [64]); however Schlick was initially impressed that in Reichenbach's approach one could avoid the use not just of rods, but also clocks (Schlick to Reichenbach, Jan. 27, 1922; [64]).

Reichenbach was moved by the conviction that complicated structures such as rods and clocks should be introduced at "the *end* of a physical theory, not at its beginning", since a "knowledge of their mechanisms presupposes a knowledge of the physical laws" ([46, 365]; tr. [56, 41]). Einstein also appreciated Reichenbach's efforts in this area (Einstein to Reichenbach, Mar. 27, 1922; [14, Vol. 13, Doc. 119]). However, on May 10, 1922, Hermann Weyl alerted Reichenbach of major shortcomings in his approach: the class of inertial frames could not be singled out by using light rays alone [44, 015-68-02]. Reichenbach only papered over the cracks of Weyl's objection [59] in the final version of his axiomatization, which was finished in March of 1923. After some difficulties finding a publisher, Reichenbach's monograph came out the following year [47].

However, the book received a lukewarm, if even hostile reception. Schlick's student Edgar Zilsel reviewed the book positively for the *Die Naturwissenschaften* (Schlick to Reichenbach, May 7, 25; [64]). However, he seemed to have considered Reichenbach's work simply a mathematically refined version of Schlick's conventionalism [81]. However, it was in particular, a dismissive review by Weyl [79], with whom Reichenbach had been on good terms, which clearly struck a hard blow. Reichenbach would still remember the episode with bitterness a decade later (Reichenbach to Einstein, Apr. 12, 1936; [9, 10-107]). Thus it is unsurprising that Reichenbach rushed to defend his work. On July 28, 1925 the *Zeitschrift für Physik* received a paper from him, which was meant to clarify some technical issues with his project, and more importantly to plea for its philosophical relevance, to "expand upon those consequences which are especially important for physics" ([48, 32]; tr. [57, 171]). The paper is relatively self-contained and provides a simplified presentation of Reichenbach's axiomatization, which we can roughly follow here [see 58, sec. 4.4.2].

As is well known, Reichenbach's axiomatization differs from the typical "deductive axiomatization" championed by, e.g., David Hilbert [3]. In the latter, one sets an abstract general principle as an axiom, such as a variational principle (see [47, 2]). Reichenbach, in contrast, put forward a "constructive axiomatization". As axioms he set empirical assertions capable of experimental verification. Reichenbach then derived the entire theory from them by integrating some additional conceptual elements, i.e., *definitions*. The definitions, a type he called "coordinative definitions", are arbitrary, thus neither true nor false. Einstein's famous definition of simultaneity is, in Reichenbach's view, a definition of this type. It amounts to the stipulation that when light signals are sent from a source to a mirror at relative rest and back again, the one-way time is $\epsilon = 1/2$ of the round-trip time. In principle, however, every value $0 < \epsilon < 1$ can be chosen.

Reichenbach's "constructive axiomatization" consisted of ten axioms, five concerning the behavior of light (*Lichtaxiome* or light axioms, I-IV), and five concerning rods and clocks (*Körperaxiome* or matter axioms, VI-X).

- The *light axioms* define the equality of spatial and temporal distances for individual frames (*Lichtgeometrie* or light-geometry) using light rays alone. In Reichenbach's view the light axioms in special relativity do not differ from those in classical theory except for the assertion that the velocity of light is the velocity's upper limit. The relativistic light-geometry claims that light propagates in spherical waves in any uniformly moving system, whereas in the the classical light-geometry light propagates in spherical waves only in the ether system. The difference depends on the choice of ϵ, which is in principle arbitrary.

- The *matter axioms* postulate that material systems used as rods and clocks behave in accordance with the light geometry (*Körpergeometrie* or matter-geometry). Space distances and time intervals that are light-geometrically equal will also turn out to be equal if measured, respectively, with rigid rods and ideal clocks. Thus the content of special relativity can be expressed by saying that rods and clocks behave according to the relativistic light-geometry and not according to the classical one. In other terms, the Lorentz transformation, which leaves the spherical propagation of light invariant, turns out to be the transformation for measuring rods and clocks. If rods transformed according to the Galileian transformations, distances traveled by light in equal times would in general not be equal if measured by rods.

Thus, in Reichenbach's axiomatization, the matter axioms contained the part of relativity theory that can be tested empirically. The only relativistic axiom that was actually put to experimental verification, according to Reichenbach, was Axiom VIII: "Two intervals which are equal when measured by rigid rods, are also light-geometrically equal" ([47, 69]; tr. [54, 89]). This amounts to nothing but an abstract formulation of the Michelson-Morley experiment. In an article published some months later [49], Reichenbach provided a good schematic description of Michelson's experimental setting (fig. 1), which we can roughly follow here.

As is well known, the essential idea of the experiment was to split a beam of light at O, allow the two resultant beams to travel along two rigid rods placed at a right angle, with two mirrors M_1 and M_2 at the end points of the arm where the beams will be reflected back. The light beams are recombined at O to produce a series of interference fringes. What has to be experimentally tested is whether the travel time of a light signal going back and forth between two mirrors depends on whether the arms are parallel or perpendicular to the direction of motion. If the travel time

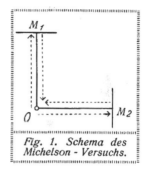

Fig. 1. Schema des Michelson - Versuchs.

Figure 1: Reichenbach's stylized Michelson-Morley apparatus [from 50, labels have been changed]

along the two paths should change when the instrument is rotated, one should observe a shift of the interference fringes ([50, 325]; tr. [57, 195f.]).

According to the ether theory, the two light beams would return at the same time only if the apparatus is at rest with respect to the ether; but since the apparatus moves along with the Earth through space, the theory demands a deviation: the ray reaching M_2 must return slightly later, producing a change in the interference pattern. Thus, by measuring interference patterns, one could determine the state of motion of the Earth relative to the ether. In the 1880s Albert A. Michelson [32], later with the assistance of Edward W. Morley ([33]), showed that "in spite of the extreme precision of the measurement, there is no difference in the time to traverse either arm of the apparatus" ([50, 326]; tr. [57, 195]). At the turn of the century, "Morley and Miller [[38, 39]] replicated this negative result in spite of the renewed increase in precision" ([50, 326]; tr. [57, 195; translation modified]).

How does one explain this negative result? Reichenbach told the following well-known story: "Lorentz in Leyden presented his explanation that assumed that all rigid bodies moving in opposition to the ether undergo a contraction" ([50, 325]; tr. [57, 197]). Lorentz justified the contraction hypothesis on the grounds that the molecular forces that hold material bodies together are of electromagnetic nature and are affected by translational motion. On the contrary, "in 1905, a more basic explanation was proposed by A. Einstein in which these contractions occur as a result of a universal principle, the principle of relativity" ([50, 326]; tr. [57, 197]). Thus Einstein could remain agnostic about the ultimate constitution of matter and only require the Lorentz covariance of all physical laws, including the unknown ones governing the material constitution of rods and clocks.

This, of course, can be cast in Reichenbach's own axiomatization, but as we shall see, this leads to some non mainstream philosophical conclusions. Axiom VIII claims that two intervals OM_1 and OM_2 (fig. 1) are equal when measured in terms of the equality of the time intervals of the round trip of the light signals ($OM_1O = OM_2O$); then they are also equal when measured by rigid rods ($M_1 = M_2$). The Michelson experiment put to the test this correspondence postulated by axiom VIII, and the result was usually accepted as correct. In particular, it received further confirmation in the work of Rudolf Tomaschek [74], an anti-relativist, who, inspired by Philipp Lenard, had repeated the experiment interferometer experiment using the light of fixed stars ([48, 39]; tr. [57, 180]).

However, Reichenbach pointed out that "[r]ecently, doubts have been raised by Dayton C. Miller, who obtained a positive result on Mount Wilson" ([48, 39]; tr. [57, 180]). If Miller's experiment were to be confirmed, then the round-trip times of the light signals along the two arms would become unequal ($OM_1O \neq OM_2O$). Therefore the matter-geometrical equality of distances would no longer coincide with the light-geometrical equality, and Axiom VIII would be disproved.

There was no agreement at the time on the reliability of Miller's results. Reichenbach, however, was quick to take advantage of the debate following the publication of Miller's paper, in order to convince the numerous skeptics of the physical implications of his axiomatization: "In this context, the axiomatization is proved to be extremely useful because it shows what particular role the Michelson experiment plays in the theory, what follows from it, and what is independent of it" ([48, 39]; tr. [57, 180]).

3 Einstein Contraction vs. Lorentz Contraction: The First Appearance of Reichenbach's Distinction

Before entering into an analysis of Miller's experiment, Reichenbach made some remarks about the philosophical interpretation of special relativity, which, I think, are quite puzzling at first glance. Reichenbach warned his readers not to subscribe uncritically to a common interpretation of special relativity:

> [W]e should examine a particular error that has crept into the understanding of the theory of relativity. It concerns the problem of Lorentz contraction and thereby leads us to the Michelson experiment. One

frequently hears the opinion expressed that in the Lorentzian explanation of the Michelson experiment the contraction of the arms of the
apparatus is an "*ad hoc* hypothesis", whereas Einstein explains it in a
most natural way, namely, as a result of the relativization of the concept of simultaneity. But this is false. *The relativity of simultaneity
has nothing to do with length contraction in the Michelson experiment.*
That this opinion is false already follows from the fact that the contraction of one of the arms of the apparatus occurs precisely in the system
in which the apparatus is at rest ([48, 43]; tr. [57, 187; my emphasis]).

As we have mentioned, in order to explain the negative result of the
Michelson experiment, Lorentz made the assumption that one arm of the
apparatus is *contracted* by the amount $\sqrt{1 - v^2/c^2}$ when it moves relative
to the ether. The theoretical asymmetry between the ether frame and those
moving with respect to it is hidden from observation by a sort of universal
conspiracy of nature. Einstein, on the contrary, considered both arms *equally
long*, if measured at relative rest in the rest system, but one arm would appear contracted by the factor $\sqrt{1 - v^2/c^2}$ if measured from a moving system.
In this way, the theoretical symmetry between the rest and moving system
is reestablished. This of course was the consequence of the fact that the
definition of the simultaneity of distant clocks using light signals is frame dependent. As any length measurement requires that both ends of the rod be
measured at the same time, two observers in relative motion refer to something different when they talk about the length of the arm of the apparatus.

Reichenbach, however, explicitly rejected this standard interpretation,
which he himself had defended not much earlier [46]. Let's take a look at
Reichenbach's own explanation:

That this opinion is false already follows from the fact that the contraction of one of the arms of the apparatus occurs precisely in the
system in which the apparatus is at rest. The "Einstein contraction"
only explains that the arm is *shortened if it is measured from a different
system*. But that does not explain the Michelson experiment. [The lat-
ter] proves that the rod lying in the direction of motion *is shorter when
measured in the rest system than it should be according to the classical
theory*. [. . .] [T]he Einsteinian theory, as well as Lorentz's, differs from
the classical theory in asserting a measurably different effect on rigid
rods that *has nothing to do with the definition of simultaneity* ([48,
43–44]; tr. [57, 187–188; translation modified; my emphasis]).

Let's assume that there is a special ether system at absolute rest in which
there are two equally long rigid rods, one of which behaves according to classical theory and the other to Einstein's theory; if we set the system in motion,
then the two rods would cease to be equally long provided that they lie along
the direction of the motion. The Lorentz-Einstein rod would be shorter than
the classical rod. The difference could in principle be measured in the moving
system itself as the difference between the rest-lengths of the classical and
Lorentz-Einstein rods. Thus a comparison with length in another system is
not at stake.

To avoid confusion, Reichenbach suggested that it is necessary to distinguish between: (a) the *Einstein contraction*, which results from the relativity of simultaneity and compares the length of the moving rod with the length of the rod at rest in the same Lorentz-Einstein theory; and (b) the *Lorentz contraction*, which compares the length of the same rod lying in the direction of motion in different theories—classical mechanics and the Lorentz-Einstein theory. In the classical theory the coordinate length of the moving rod is expected to be just as the coordinate length of the rod at rest. In the Lorentz-Einstein theory, by contrast, the coordinate length of a moving rod is always shorter than its proper length which is the same in all inertial frames. Nevertheless, Reichenbach claims, the proper length of a rod in motion can still be said to be shorter than the length that a classical rod would have if both were measured at relative rest in the moving frame [15].

Reichenbach uses the following notation. Let's call l a rod that behaves according to the Lorentz-Einstein theory and L a rod that behaves according to the classical theory. Let's label K the rest system and K' the moving system. The rest lengths of l and L in K are equal, or, as Reichenbach put it, $l_K^K = L_K^K$. In his notation the upper index refers to the system in which the rod is measured and the lower refers to the one in which the rod is at rest. Thus the rest-length of a moving rod from the perspective of a co-moving frame, in other terms its proper length, is $l_{K'}^{K'}$; the length of a rest-rod from the perspective of the moving frame $l_K^{K'}$ is its coordinate length.

Now let's consider what happens in the system K' that is in uniform motion with respect to K. The *Lorentz contraction* is concerned with the ratio $l_{K'}^{K'} : L_{K'}^{K'}$, whereas the Einstein contraction is concerned with the ratio $l_{K'}^K : l_K^K$. That is, the Lorentz contraction compares the behavior of the same rod in the Lorentz-Einstein and classical theories in the same inertial system K'. The *Einstein contraction* compares the behavior of two rods in the Lorentz-Einstein theory in different inertial systems, K and K'. The Einstein contraction depends on the relativity simultaneity and "is related to the comparison of *different magnitudes* within the *same theory*" ([48, 44]; tr. [57, 188]) (coordinate and proper length). An analogous example would be the annual parallax, the difference in the position of a star as seen from two different extremes of the Earth's orbit. The Lorentz contraction is related to "the behavior of the *same magnitudes* according to *different theories*" ([48, 45]; tr. [57, 188]) (the classical and relativistic proper length). An analogy to this would be the difference between the gravitational light deflection, which in general relativity is twice the Newtonian value.

According to Reichenbach, only the Lorentz contraction is at stake in the Michelson experiment, not the Einstein contraction: "It just happens that both contractions depend upon the same measurement factor $\sqrt{1 - v^2/c^2}$, and this is probably the reason why they are always confused with one another" ([48, 46]; tr. [57, 189]). Reichenbach shows that the equality $l_{K'}^{K'} : L_{K'}^{K'} = l_{K'}^K : l_K^K$ is simply a consequence of the linearity of transformation. However, one should not miss the deep conceptual difference between the two contractions, which is hidden behind the coincidental numerical equality of the two factors ([48, 45f.]; tr. [57, 189f.]). Thus in Reichenbach's view, it is not advisable to use the same expression "contraction" in both cases.

It is this very expression "contraction" that Reichenbach finds misleading. It implies that physical objects satisfy the classical theory without a cause, so that one must search for a cause for deviations from the correct behavior. However, there is no reason to see the classical theory as natural, and the Lorentz-Einstein theory as a distortion. The problem of causality should be posed in a different form; one must explain why measuring rods and clocks conform to a certain set of transformations defined in terms of the light-geometry and not to a different one. This causal problem is the same whether the rods and clocks behave according to the relativistic or classical transformation. In his 1924 monograph ([47, 70–71]; tr. [54, 90-91]), we already find Reichenbach using Weyl's expression "adjustment" as a good way to express this peculiar form of causality.

As Reichenbach explains, Weyl [78] had introduced the expression "adjustment" to account for the surprising behavior of the physical systems, such as atoms, that we use as rods and clocks. It cannot be a coincidence that atoms of the same type always have the same Bohr radius, independent of what happened to them in the past; this fact suggests that each time, they "adjust" anew to a certain equilibrium value, rather then "preserve" it. The analogy with special relativity seems to be the following: "*Einstein's* idea can be formulated as meaning that *light geometry and matter geometry are identical*" ([47, 11]; tr. [54, 14]). It is an odd coincidence that any physical system we use as a rod—whether it is made of steel, wood, etc.—always measures at equal lengths that are light-geometrically equal. "Light is a much simpler physical object than a material rod, and, when searching for a relation between the two, it should be initially supposed that it would not correspond to so ideal a scheme as the posited matter axioms" ([48, 47-48]; tr. [57, 95]). This coincidence cries out for an explanation. However, the explanation should not account for the *divergence* from an alleged correct behavior, but a for the *convergence* toward a non-trivial one:

> The word adjustment, first used in this way by Weyl, is a very good characterization of the problem. [...] [A]ll metrical relations between material objects, including the observed fact of the Michelson experiment, must therefore be explained in terms of the particular way in which rigid rods adjust to the movement of light. Of course, the answer can only arise from a detailed theory of matter about which we have not the least idea [...] [.] The word "adjustment" here thus only means a problem without providing an answer; the relevant fact is strictly formulated in the matter axioms without using the word "adjustment". Once we have this theory of matter, we can explain the metrical behavior of material objects; but at present the explanation from Einstein's theory is as poor as Lorentz's or the classical terminology ([48, 46–47]; tr. [57, 191]).

According to Reichenbach, the difference between Lorentz and Einstein's theories is not in their empirical content, but in the types of explanation they provide. Both assert the facts encoded in axiom VIII, whereas the classical theory denies them. However, Lorentz's theory assumes the classical behavior of rods as "self-evident", so that any *deviation* from these relations must have a cause. Einstein's theory *renounces* the explanation and axiomatically *defines* two rods as equal if they behave in accordance with the Michelson experiment. "The superiority of Einstein's theory lies in the recognition of the epistemological legitimacy of this procedure" ([52, 233]; tr. [53, 202]). However, according to Reichenbach, Einstein's agnosticism is unsatisfying. Without a suitable theory of matter describing those physical systems we happen to use as rods and clocks, Einstein's account of the metrical behavior of material objects is "just as poor as Lorentz's" ([48, 43]; tr. [57, 187]).

4   Reichenbach on Miller's Experiment

After Reichenbach clarified the distinction between Lorentz and Einstein contraction, he could proceed further to show what would happen if a Michelson-Morley-type experiment gave a positive result. As we have mentioned, in those years, raising this issue was more than just a mental exercise. A few months before the publication of Reichenbach's paper, Miller had published the result of his Mount Wilson experiments [35].

Miller had begun working with Morley on the detection of the ether drift twenty years before, and they had published a null result [38, 39, 40]. The observations were made on slightly elevated locations and had indicated the occurrence of a small displacement. Miller conjectured it was significant. He remounted the Morley-Miller apparatus at a higher elevation at the Mount Wilson Observatory in 1921, moved it to Cleveland in 1922 [34], then back to Mount Wilson at 1734 metres above sea level in 1924. There Miller found a positive displacement of the interference fringes, approximately 10km/s, such as would be produced by a relative motion of the Earth and the ether (instead of the nearly 30 km/s expected). This suggested a partial drag of the ether by the Earth, which decreases with altitude [35].

Miller's result immediately sparked considerable debate in the physics community. On May 23, 1926 the polish-born physicist Ludwik Silberstein (author of one the first special-relativistic textbooks) published an announce in *Nature* claiming that Miller's result refute special relativity and support Stokes-Planck theory based on the idea on a compressible ether [67]. Arthur Stanley Eddington the major relativist in the English speaking world, replied on June 6, 1925 attacking Miller's result: if the ether drag depended on altitude, then also astronomical observations should be different on Mount Wilson respect to that on sea level [8].

A few week later, at the end of July, Reichenbach was the first "philosopher" to attempt to participate to the debate, that he clearly saw as a good opportunity to convince the numerous skeptics of the validity of his axiomatic. Reichenbach's reaction reveals that the implications of Einstein's experiment were clearly non-mainstream:

> Now we can also address the question what would change in the theory
> of relativity if Miller's experiment were held to prove that the hith-

erto negative result of the Michelson experiment is in principle wrong. *Nothing would change* in Einstein's theory of time as it has nothing to do with the Michelson experiment. Also nothing would change with the light geometry; it remains in any case a possible definition for the spacetime metric and probably a much better and more accurate one than the geometry of rigid rods and natural clocks. *But what would change is our knowledge about the adjustments of material things to the light geometry.* With respect to the matter axioms, as far as they differ from the classical theory, the Michelson experiment is the only one that has been confirmed. If this should be refuted, one has to develop a more complex view of the relationship between material objects and the light geometry ([48, 47]; tr. [57, 192; my emphasis]).

In Reichenbach's axiomatization, the Michelson result is summarized in Axiom VIII. Thus, in the event that Miller's experimental results were not spurious, only this axiom would change. The principle of the constancy of the speed of light could be maintained, since it depends on a definition; thus one could construct a "light geometry" using light signals but employing no rigid rods. "From this perspective, the Michelson experiment serves only as a bridge [Verbindungsgliedes] between the light geometry and the geometry of rigid rods" ([48, 327-328]; tr. [57, 203]). If the experiment were rejected, this would only mean that "rigid rods do not after all possess the preferred properties that Einstein still attributes to them" ([48, 328]; tr. [57, 203]). However, this would not imply a return to the old ether theory, but only a change in the matter axioms. Whereas the light axioms are completely certain, the matter axioms make statements about very complicated material structures. In Reichenbach's view they might have the "the validity of a first-order approximation in the same way that the ideal gas law cannot be maintained if the accuracy is increased" ([48, 48]; tr. [57, 192]).

5 Schlick's Reaction to Reichenbach's Paper

Einstein drew very different consequences from Miller's experiment. On June 26, 1925, two days before Reichenbach's paper was submitted to the *Zeitschrift für Physik*, Edwin E. Slosson, the executive editor and director of the *Science Service* asked Einstein for a comment. Einstein's stance towards the issue at that time was clearly expressed in a letter that he sent a few days later to Robert A. Millikan, Caltech's "chairman of the executive council": if the Miller's result turned out to be correct, Einstein wrote, then "the whole theory of relativity would go down like a house of cards" (Einstein to Millikan, July 13, 1925; [9, 17-357]). A few days later, Einstein sent a very similar statement to Slosson which was published on *Science* on July 31, 1925. Against Ludwik Silberstein's [67] claim the Miller's results support the Planck-Stoke theory, Einstein argued that they would mean a return to Lorentz's theory: "No theory exists outside of the theory of relativity and the similar Lorentz theory which, except for the Miller experiment, explains all the known phenomena up to date" [68].

In general, however, physicists expressed serious doubts, including Millikan and his group at Caltech (Epstein to Einstein, July 25, 1925; [9, 10-565]). Einstein himself was not convinced that Miller's experiment was reliable. As he wrote to Paul Ehrenfest in August 1925: "In the deep of my soul, I take no stock in Miller-experiment at all, but I cannot say it loudly" (Einstein to Ehrenfest, Aug. 18, 1925; [9, 10-108]). Einstein's skepticism, just like Eddington's, was motivated by the fact that a difference in height between Cleveland und Mount Wilson was not enough to explain Miller's results. In September 1925, Einstein suggested that a temperature difference (Einstein to Ehrenfest, Sep. 18, 1925; [9, 10-111]; Einstein to Piccard, Sep. 21, 1925; [9, 19-211]) could be the source of Miller's findings. He communicated his guess to Miller, (cf. (Miller to Einstein, May 20, 1926; [9, 17-274])), who by the end of the year, published a more detailed account of the experiment [36].

In this context, it is not surprising that Reichenbach's entirely different attitude towards Miller's results could appear puzzling. At the end of 1925, Schlick expressed disconcert in his correspondence with Einstein (Schlick to Einstein, Dec. 26, 1925; [9, 21-591]). "Mr. Reichenbach" he wrote "has recently published a paper 'Über die physikalischen Konsequenzen der relativistischen Axiomatik' in the *Zeitschrift für Physik*". Schlick was eager to know Einstein's opinion, since the paper "quite clearly shows the limit of the axiomatic method" (Schlick to Einstein, Dec. 26, 1925; [9, 21-591]).

Schlick was understandably confused by Reichenbach's claim that the Lorentz contraction is not *ad hoc*; after all, this had been Schlick's reading of special relativity for at least a decade [60]. Reichenbach's paper, by attacking a not further identified mainstream interpretation of special relativity, seems to have attacked more or less explicitly Schlick's interpretation. Schlick had still recently defended it at the 1922 Leipzig meeting *Gesellschaft Deutscher Naturforscher und Ärzte* [61]. As is well known, Schlick considered Lorentz and Einstein's theory as empirically equivalent, that is equally consistent with all experiential data. However, whereas Lorentz theory introduces compensatory contractions and retardations, Einstein's theory avoid the introduction of redundant theoretical structure, and it is thus preferable. Lorentz's theory, in Schlick's view, is analogous to the attempt of saving Euclidean geometry, by adding a force field deforming all measuring instruments. The difference is not a matter of truth, but a matter of simplicity. As we as seen, within the Schlick's circle, Reichenbach's axiomatic was initially received as a more sophisticated presentation of this point of view. However, Schlick now realized that Reichenbach had something very different in mind:

> The considerations on p. 43 [. . .] show in my opinion that his axiomatic cannot distinguish between special relativity and Lorentz's theory (with the contraction hypothesis). This seems to me trivial since the equations are the same. The real difference between the two theories is a philosophical one and cannot be grasped in the logical way of the axiomatic. This difference can be aptly expressed through the parlance that Reichenbach rejects: Lorentz had introduced an *ad hoc* hypothesis. Even if, from a logical point of view, spec. rel. theory

must introduce as many assumptions as Lorentz's theory. In the first case they naturally fit in the framework of the relativity thought and the contraction hypothesis is psychologically not *ad hoc*. On the contrary, in the case of Lorentz-Fitzgerald [the contraction] appears as an element added *ad hoc* (Schlick to Einstein, Dec. 27, 1925; [9, 21-591]).

The limits of Reichenbach's approach, according to Schlick, emerge even more clearly from Reichenbach's reaction to Miller's experiment. If the experiment were to be confirmed, Schlick argued, the universal conspiracy of nature that hides the ether from detection would be broken, and we had to return to the ether theory. By contrast, Reichenbach's Axiomatic method is incapable of grasping the difference between the two cases:

> Also the paper's remark—about the possible interpretation of Miller's experiments—does not seem to be to grasp the philosophical key point. If the experiments would really prove (and this is surely not the case), that a particular direction (that of the "aether wind") were privileged, one would certainly abandon the relativistic physics; even if it were possible to keep relativity through the assumption of certain "matter axioms", one would certainly not take this path. Against this, the axiomatic consideration remains indifferent. In the strict sense, one cannot speak of physical consequences of the axiomatic. The question seems to me philosophically relevant. I would be deeply grateful if you could tell me in a few lines if I'm right (Schlick to Einstein, Dec. 27, 1925; [9, 21-591]).

Unfortunately, Einstein did not comply Schlick's request and he probably never read Reichenbach's paper. However, as we have seen, there is no doubt that Einstein's position was closer to that of Schlick.

Einstein had defined the Lorentz contraction hypothesis as *ad hoc* in the past [10, 11], provoking Lorentz's reaction (Lorentz to Einstein, Jan. 23, 1915; [14, Vol. 8, Doc. 43]). Also concerning Miller's experiment, Einstein clearly agreed with Schlick [20, 361f.]). On January 19, 1926 a brief, but unequivocal statement of Einstein was published in the at that time most influential German newspaper, the *Vossiche Zeitung*: "If the results of Miller's experiments should indeed be confirmed, the relativity theory could not be upheld" [12]. Einstein was clearly convinced the Miller's result were probably spurious, and he was ready to put money where its mouth was, as he explicitly phrases it [12]. However, he had no doubt that, if Miller's experiments turned out to be correct, the relativity principle would have to be abandoned entirely [21]. This hardly surprising. In Einstein's view, the construction of a device detecting the ether-drift would have been comparable to the construction of a *perpetuum mobile* of the second kind þ. The systematic failure of the ether drift experiments provided the empirical support the relativity postulate, with which the entire theory stands or falls.

A few months later, Reichenbach submitted a popular paper on Miller's experiment entitled 'Ist die Relativitätstheorie widerlegt?' which appeared in the weekly magazine *Die Umschau* on April 24, 1926. By that time, Reichenbach was fully aware of what "Einstein himself has recently said in the newspapers,"; however, he saw no reason to abandon his "less radical opinion" ([50, 327]; tr. [57, 202]), namely that *"Miller's result in no way affects the philosophical consequences of the theory of relativity"* ([50, 328]; tr. [57, 203; my emphasis]). It would only imply a change in our knowledge of the physical mechanism governing rods and clocks [19, 308].

Reichenbach too expressed some doubts about the correctness of Miller's experiment. The curve determined by Miller deviates from the expected symmetry with respect to the horizontal axis (fig. 2). J. Weber [77] showed that Miller's figure omits some quite problematic measurement data without explanation. It was unlikely that on Mount Wilson, which after all is merely 0.03% of the Earth's radius, one would detect an ether wind 1/3 less than expected [73]. Moreover, Rudolf Tomaschek [75, 76], an anti-relativist, performed the so-called Röntgen-Eichenwald and Trouton-Noble experiments and obtained negative results on the *Jungfraujoch* in the Bernese Alps. Thus, it would have been interesting to replicate a Michelson-type experiment on the *Jungfraujoch* before drawing further conclusions.

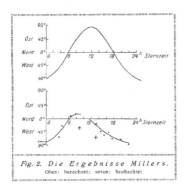

Fig. 2. *Die Ergebnisse Millers.*
Oben: berechnet; unten: beobachtet.

Figure 2: The Results of Miller's Experiment [from 50]

However, the philosophical point of course lied elsewhere: *"What then does the theory of relativity have to infer from Miller's experiment?"* ([50, 327]; tr. [57, 202]). Concerning this point, Reichenbach was not afraid to express an opinion that radically differed from that of Einstein. Somehow anticipating his later famous distinction between the context of discovery and the context of justification, Reichenbach claimed that "[t]he Michelson experiment, of course, played a crucial role in the *historical development* of the theory" ([50, 327]; tr. [57, 202; my emphasis])[1]; however, according to Reichenbach, "it does not occupy this same significant place in the relativistic theory's *logical structure*" ([50, 327]; tr. [57, 202; my emphasis]).

The logical structure of the theory was of course expressed by Reichenbach's own axiomatization:

> Under the ten axioms of the theory of relativity as I have laid them out, i.e., its ten most basic empirical propositions, there is only one that entails the Michelson result; it is only this axiom then that is thereby threatened. The principle of the constancy of the speed of light could be maintained in a more limited form even if the Michelson experiment's negative result were overturned. One could construct a "light geometry" using light signals but employing no rigid rods to maintain a metrical understanding of the world and allow the previous

---

[1] This claim is actually problematic [cf., e.g., 70, 5].

formulation of all physical laws. From this perspective, the Michelson experiment serves only as a bridge between the light geometry and the geometry of rigid rods. Should this connection be lost, this would only mean that rigid rods do not after all possess the preferred properties that Einstein still attributes to them. This would not mean a return to the old aether theory, but rather a step towards the renunciation of a preferred system of measurement in nature ([50, 327]; tr. [57, 203]).

Even after a positive result of the aether-drift experiments, one could still maintain the light postulate, the claim that the motion of light is a spherical wave for any uniformly moving system, by adopting a non-standard definition of simultaneity. However equal lengths measured by rods and clocks would not be equal to equal lengths measured by light rays. Thus rods and clocks would adjust to the classical light geometry, and not to relativistic one.

Reichenbach, in his role of the defender of Einstein's theory, seems to have become more royal than the king. What Reichenbach probably meant, is that the real philosophical achievement of the theory is to have revealed the logical structure which was ultimately encoded in his axiomatization. Special relativity has disentangled conceptual elements, which in the classical theory appeared to be confused. There are empirical statements that serve as axioms (which can be true or false independently of it) and definitions (which are neither true nor false). Light axioms can be true or false, but simultaneity is a definition. In this sense, the Galilei and Lorentz transformations are neither true nor false, since both agree on the light axioms and are different because of the definition of simultaneity [50]. The matter axiom can be true or false since it is a matter of fact whether they adapt to Lorentz or the Galilei transformations. This entire axiomatic structure, however, "is independent of specific, physical observations" ([50, 203]; tr. [57, 328]). In other terms, Miller's experiment, if confirmed, would disprove axiom VIII, but would not engender the logical structure of Reichenbach's axiomatization. It is true that such logical structure has been derived "from a particular physical theory", namely special relativity, but the latter has "given rise to philosophical insights which no longer belong to the realm of physics but rather to the philosophy of nature" ([50, 204]; tr. [57, 328]).

In the meantime, Reichenbach thanks Max Planck's and and Max von Laue's support, had obtained the chair for natural philosophy in Berlin (Reichenbach to Schlick, July 2, 1926; [64]; Schlick to Reichenbach, July 5, 1926; [64]). Reichenbach clearly continued to consider the difference of this two types of contractions important. An entry in Rudolf Carnap diaries, who met Reichenbach in the Berlin region in September 1926, reads: "He explained me the difference between Lorentz and Einstein contraction" [2, 025-72-05]. Carnap had just joined the Schlick circle in Vienna. The philosophical disagreement between the Schlick-Circle and Reichenbach was deepening and went beyond the specific issue of the philosophical interpretation of special relativity. Reichenbach was convinced that the philosophy of nature should tackle metaphysical questions, such as that of the reality of the external world or the human freedom [51]. By contrast, Schlick, influenced by the young Carnap and Ludwig Wittgenstein deemed all such metaphysical questions as non-sensical [62].

We do not know whether Reichenbach ever became aware of Schlick's negative opinion of his interpretation of special relativity. However he clearly did not made any conciliatory steps: Reichenbach's line of argument can be found again in the *Deutsche Literaturzeitung* [52], which, as a letter from Reichenbach to Schlick reveals, was already finished at the end of 1926 (Reichenbach to Schlick, Dec. 6, 1926; [64]), though Reichenbach was only able to find a publisher months later. When the book was finally published that the beginning of 1928 the importance of Miller's experiment was fading away especially in Germany. According to Reichenbach, "Michelson experiment has been confirmed to a very high degree" and he considered "this matter closed" ([52, 225]; tr. [53, 195]). A similar opinion was expressed by Einstein in 1927 [13]. However Reichenbach still considered important to address "erroneous interpretations in the usual discussions on relativity" ([52, 225]; tr. [53, 195]) that the discussion of Miller's experiment had revealed. In § 31 Reichenbach almost literally repeats the content of the 1925 paper we started with, as if the discussion of Miller's experiment would force him to clarify an central issue of his interpretation of special relativity. However, he also made some important clarifications. In particular he introduced his distinction between Lorentz and Einstein contraction in terms of Minkowski diagrams.

6 Lorentz Contraction vs. Einstein Contraction in Terms of Minkowski Diagrams

6.1 Minkowski Spacetime as a Graphical Representation

Reichenbach had a somewhat deflationary attitude towards Minkowski spacetime. He viewed it as nothing but a "graphical representation", an expression he probably borrowed from Arthur Stanley Eddington [7]. Reichenbach defines "graphical representations" as structural analogies between different physical systems (e.g., compressed gases, electrical phenomena, mechanical forces, rigid bodies and light rays, etc.), which are realizations of the same conceptual system (e.g., the axioms of Euclidean geometry) ([52, 123ff]; tr. [53, 101ff]).

In the case of Minkowski spacetime, if "we speak of a geometrization of physical events, this phrase should not be understood in some mysterious sense; it refers to the identity of types of *structure* and not to the *identity of the coordinated physical elements*" ([52, 220]; tr. [53, 190]). By asserting that measuring rods, clocks, and light rays behave according to the relations of congruence of the indefinite metric, Minkowski spacetime provides the geometrical representation of the light and matter axioms. When in fig. 3 we "symbolize", say, the motion of rod OS with its rotation of the segment around O, "we only g[ive] a graphical representation, which means that the logical structure [*Beziehungsgefüge*] exhibited by the rods [...] [c]an also be realized by the spacetime manifold" ([52, 220]; tr. [53, 190]).

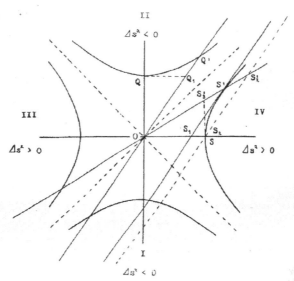

Zur Realisierung der indefiniten Metrik durch Uhren und Maßstäbe.

Figure 3: Realization of the indefinite metric by means of light rays, clocks and measuring rods; from [52, 215]; the label S_2 has been added

As Reichenbach put it, in Minkowski spacetime a number Δs is coordinated to the coordinate differences $\Delta x_1, \ldots, \Delta x_4$ by means of the fundamental metrical formula $\Delta s^2 = \Delta x_1^2 + \Delta x_2^2 + \Delta x_3^2 - \Delta x_4^2$. The minus sign in the rule for computing real distances from coordinate distances is responsible for all the differences between the Minkowski and Euclidean geometry. The lines which are at a constant distance from the origin at Δs satisfy the equation $\Delta x^2 = \Delta x_1^2 - \Delta x_4^2$ rather than $\Delta x^2 = \Delta x_1^2 + \Delta x_4^2$. The contour lines are hyperbolae, and not circles like in Euclidean geometry. The four-hyperbolae in fig. 3, like the unit circle in Euclidean geometry, are the set of all points at $\Delta s = 1$ distance from O. The hyperbola $\Delta x^2 = 0$ degenerates into the two asymptotes, which group all other events in spacetime into three different classes of intervals characterized by the sign of the quantity Δs^2. As one might expect, Reichenbach's next step is to find measuring instruments that behave like the indefinite type of metric, just like the behavior of the rods correspond to the definite one.

The physical realization of the negative Δs^2 is a physical object that satisfies the relations of congruence defined by the hyperbolas of quadrants I and II. The realization of the positive Δs^2 is a physical object that satisfies the relations of congruence defined by the hyperbolas of quadrants III and IV. The first is called a time-like interval $\Delta s^2 = -1$ and is realized by the proper time of a clock. The rotation of the interval OQ into the position OQ' represents a moving clock. $\Delta s^2 = 1$ is the space-like interval and is realized by the proper length of a rod. The rotation of interval OS into OS' sets the rod into motion. Light rays realize $\Delta s^2 = 0$, the limiting velocity, which cannot be reached but only approached arbitrarily closely. Otherwise rods and clocks behave by following the hyperbolic contour lines.

As Reichenbach rightly notices, there is a deep disanalogy between clocks and rods. Clocks are intrinsically four-dimensional measuring instruments, since they measure distances between two events. Measuring rods, on the other hand, are three-dimensional measuring instruments; they can be treated as four-dimensional instruments, if events are produced at their endpoints according to the appropriate definition of simultaneity ([52, 217]; tr. [53, 187]). It is from this difference that all difficulties arise concerning the behavior of rods.

6.2 Lorentz vs. Einstein Contraction in Minkowski Spacetime

According to Reichenbach, resorting to the geometrical representation in fig. 3, one can easily recognize the difference between Lorentz and Einstein contraction. In Minkowski spacetime the history of a uniformly moving unit rod is represented by a world-strip bounded by the parallel world-lines of the rod's endpoints. In the Lorentz-Einstein theory—keeping in mind that the points on the hyperbolas are at distance 1 from O—the moving rod is represented by the narrower strip between the world-lines OQ' and S_1S'; according to the classical theory it is represented by the wider strip between OQ' and SS_2'. The Einstein contraction maintains that the moving rod $OS' = 1$ looks shorter from the perspective of the rest frame (OS_1) than the proper length of the rod $OS = 1$ ($l_{K'}^K < l_K^K$). The Lorentz contraction refers to the fact that the classical length OS_2' would be longer than the relativistic proper length $OS' = 1$, if both were measured in the same moving frame ($l_{K'}^{K'} < L_{K'}^{K'}$) ([52, 225]; tr. [53, 195]).

"This assertion of the theory of relativity is based mainly on the Michelson experiment" ([52, 226]; tr. [53, 195]). The Michelson experiment proves that material rods satisfy the light-geometrical definition of congruence in all inertial systems. The matter-geometrical equality of distances happens to coincide with the light-geometrical equality. In other terms, rods set in motion behave according to the hyperbolic contour lines in quadrant IV. Consider again fig. 1. OM_1 and OM_2 are regarded as equally long if light rays need equal time when they are sent back and forth along OM_1O and OM_2O. The negative result of the Michelson-Morley experiment establishes that if OM_1 and OM_2 are equally long if measured through light signals (in terms of the absence of interference fringes), then small measuring rods that are placed along the arms also mark off an equal number of segments on both arms:

$$(OM_1O = OM_2O) \rightarrow (OM_1 = OM_2) \tag{1}$$

According to the classical theory, the implication (1) is satisfied only in the ether frame. In all other frames moving through the ether, the rest-length of the rod oriented in the direction of motion will no longer satisfy the implication (1). If OM_1 and OM_2 are the arms of the Michelson-Morley interferometer, then the fact that the matter-geometrical equality of distances does not coincide with the light-geometrical equality is revealed empirically by the shift in the patterns of light and darkness detected by the apparatus. The Lorentz-Einstein theory claims that the implication (1) is satisfied in all frames; the light-geometrical distance always coincides with the matter-geometrical distance in all inertial systems. Light geometry is the same in both cases.

Thus, both Einstein and Lorentz's theories assume that the proper length of the arm of the Michelson apparatus lying in the direction of motion is shorter than it *would be* according to the classical theory. An objection that immediately comes to mind is that it is impossible to compare two magnitudes belonging to different theories, since there is no common standard of comparison. However, according to Reichenbach, his axiomatization provides the common standard: "In this case, the *tertium comparationis* is light, which in terms of light-geometrical definitions supplies a standard to which the rods of the different theories can be compared" ([52, 197]; tr. [53, 228]). The hyperbola in quadrant IV (fig. 3) defines the distance 1 from O. Rods in motion, that is, rotated around O, follow the hyperbolas in the Lorentz-Einstein theory, but they do not do so in the classical theory. In this way, in Reichenbach's view, it is possible to compare rods I and L, though only one of them has an actual physical existence.

Thus, the Lorentz contraction is a *real difference*, just as the pressure of gas is really lower according to van der Waals equation than it would be according to the ideal gas equation. This does not mean that Einstein contraction is "apparent". Reichenbach prefers to speak of a *metrogenic* or (since motion is implied) metrokinematic difference: it depends on the fact that two observers in relative motion measure two different three-dimensional cross-sections of the world-strip of the rod; thus it is a "perspectival difference" ([52, 228–229]; tr. [53, 197]). The Michelson experiment implies a real difference between the classical theory and Einstein's theory, as well as between the classical theory and Lorentz's theory. But *there is no difference between Einstein's and Lorentz's theories*: "The concept of simultaneity does not enter into this problem at all" ([52, 229]; tr. [53, 198]).

In geometrical terms, the Einstein contraction compares the width of different three-dimensional simultaneity cross-sections of the same relativistic world-strip (it is a perspectival difference); the Lorentz contraction compares the width of same three-dimensional cross-section of different world-strips (it is a real difference). fig. 3 reflects the fact that in the classical theory there is neither a Einstein nor a Lorentz contraction ($OS = OS_2 = OS'_2$). In the Lorentz-Einstein theory the Einstein contraction is Lorentz contraction ($OS'_1 < OS'_2 \rightarrow OS_1 < OS$). However, the two contractions are not identical.

This is essential to understand "[w]hat is the difference between Einstein's and Lorentz's theories" ([52, 229]; tr. [53, 198]). In order to answer this question Reichenbach distinguishes between the following two statements ([52, 229]; tr. [53, 198]):

1. the length of the moving rod $l^K_{K'}$ measured from the rest frame is different from its proper length l^K_K. As is well known, the proper length is greater than any coordinate length; the difference disappears only for the co-moving observer.

2. the rest-length of the moving rod $l^{K'}_{K'}$ is different from the rest-length of another rod $L^{K'}_{K'}$ which moves with it but satisfies the classical theory. The relativistic proper length in the moving frame is shorter than the classical length would be as judged from the same frame.

In Reichenbach's view, (b) can either be "true" or "false" in both Lorentz and Einstein's theories, depending on whether one accepts, say, Michelson or Miller's experimental results. In the geometrical representation, it is indicated by the difference between the distances $l^{K'}_{K'} = OS'$ and $L^{K'}_{K'} = OS'^2$ (fig. 3). On the contrary, (a) is neither "true" nor "false"; it depends on the definition of simultaneity adopted (one can choose the standard Einsteinian definition or an alternative one). In the geometrical representation, statement (a) is equivalent to the comparison of $l^K_K = OS$ and $l^K_{K'} = OS'$. Lorentz believed that lengths mentioned in (a) are different because the lengths mentioned in (b) are: the lengths of the arms of the interferometer are equally long only in the ether frame. On the contrary, Einstein declared the lengths mentioned in (a) are equal if measured at relative rest: the proper lengths of the arms of the interferometer are the same in all inertial frames ([52, 229–230]; tr. [53, 198–199]). Reichenbach, however, made a further statement embodying the peculiarity of his approach: "*It is sometimes overlooked by proponents of the theory of relativity that statement (b) is nevertheless true*" ([52, 230]; tr. [53, 198]) in Einstein's theory as well.

Thus Einstein's theory, just like Lorentz's, implies a contraction that is independent of the relativity of simultaneity, namely, the Lorentz contraction that implies a comparison of lengths $l^{K'}_{K'} < L^{K'}_{K'}$, i.e., $OS' < OS'^2$ in the same moving frame but in different theories. In addition, however, it contains the Einstein contraction, which compares lengths in the same theory: the proper length and the coordinate length $l^K_{K'} < l^K_K$, i.e., $OS < OS_1$. As we have seen, Reichenbach maintained the opinion that the two contractions only happen to amount to the same Lorentz factor as a consequence of the linearity of the Lorentz transformations. The *numerical identity* $OS_1 : OS = OS' : OS'^2$ is coincidental and it conceals a deeper *conceptual difference*.

Reichenbach repeats the proof of this statement, as it appeared in his 1925 article ([52, 230f.]; tr. [53, 199f.]); however, in order to emphasize the difference between the two contractions, he constructs a counterexample in which an "Einstein contraction" appears but there is no "Lorentz contraction". The example is based on the possibility of using Einstein's definition of simultaneity or an alternative one ($\epsilon \neq 1/2$) in the classical theory ([52, 231]; tr. [53, 200]). Reichenbach's conclusion can be understood without entering into too much detail:

> The example [...] [m]akes it particularly clear that the Einstein contraction is a metrogenic phenomenon. In the geometrical representa-

tion this means that we may choose as the length of the rod differently directed sections through the world-strip of the rod. On the other hand, the geometrical representation of [fig. 3] shows very clearly that through the difference in the width of the strip, the Lorentz contraction indicates a difference in the actual behavior of the rod. These considerations also explain how it is possible to compare rods I and L, although only one of them is physically realized. OS is the same in both theories; the classical theory claims that the right-hand boundary of the strip parallel to OQ' must be drawn through S, whereas the new theory places the boundary along the tangent to the hyperbola which passes through S' ([52, 232]; tr. [53, 200]).

Thus it is correct to claim that the Einstein contraction does not require any physical explanation; it is a metrogenic difference between proper and coordinate length. However, Lorentz's contraction *does* cry out for such an explanation. The negative result of the Michelson experiment implies that rods of all materials invariably behave in agreement with distances measured by light rays. How can such a coincidence be explained?

As we have seen, for Reichenbach, Weyl's expression "adjustment" aptly expresses the need for an explanation, but it provides no details as to what it would look like. "The answer can of course be given only by a detailed theory of matter, of which we have not the least idea" ([52, 233]; tr. [53, 201; translation modified]). It is important to emphasize that this is not a marginal aspect of Reichenbach's philosophy. According to Reichenbach the very same problem emerges in general relativity when the non-Euclidean nature of the continuum is taken into account. In this case, too, he resorts to the expression "adjustment" and he refers his readers to the very same §31 of the book.

In general relativity measuring instruments in a gravitational field should also be considered "free from deforming forces", and not deviating from the Euclidean expected behavior; nevertheless one may still seek an explanation for why all measuring instruments happen to agree on the *same*, generally non-Euclidean, geometry. According to Reichenbach, in general relativity, too, only a theory of matter can explain this peculiar behavior of the space-time measuring instruments. In the presence of a real gravitational field it is impossible to arrange rods and clocks in a rectangular grid, just like it is impossible to "develop" a flat piece of paper around a sphere. "We know that a more detailed investigation would reveal the presence of molecular force-fields, which affect the molecules on the surface of the sphere and thus force it into a definite" ([52, 295]; tr. [53, 258]) congruence relationship.

Conclusion

The interest in Miller's results faded away in the 1930s. The two great American experimentalists of their time, Miller and Charles Edward St. John [69], discussed the issue at the April 1930 meeting of the National Academy of Sciences [30, 4.3]. St. John seemed to have won the debate, but Miller continued to defend his position in a long paper published just thereafter [37]. However, the Kennedy-Thorndike experiment [29] brought further grist to special relativity's mill, and so did the Ives-Stilwell experiment [22, 23] a few years later in spite of Charles Ives' intentions [31]. Miller results were considered inexplicable until in 1954, when Robert Shankland et al. [66], who also discussed the matter with Einstein [65], put forward an explanation of Miller's data which is usually accepted, even if not universally [30].

From the point of view of the history of philosophy of science, Miller's experiments played the role of a litmus test. Even like-minded philosophers, like Schlick and Reichenbach, turned out to have a quite different stance towards Miller's results. Schlick reacted as expected, rehearsing his conventionalist position. The difference between Lorentz and Einstein theory is not a matter of truth, but a matter of choice. The superiority of Einstein's theory depends on the fact that the theory involves less redundant elements, in particular, no *ad hoc* molecular dynamical contraction of the rods. On the contrary, Reichenbach was forced to reveal some more striking consequences of his axiomatization program. Reichenbach considered the contraction of rods as a physically "real" contraction, in the sense that in both theories rods behaves differently than in the classical theory. Einstein's and Lorentz's theories do not differ in this respect. In both cases a "molecular" explanation of the contraction is needed. If Lorentz erroneously conceived the explanation as a deviation caused by a force, Einstein proved negligent by avoiding any explanation. According to Reichenbach, this point had been misunderstood, since the *explanandum* had not been properly identified. What has to be explained is not the metrogenic and perspectival "Einstein contraction", but the physical and real "Lorentz contraction", that is the odd coincidence that measurements made by material structures always agree with measurement made by electromagnetic signals.

Reichenbach's distinction between two type of contractions is seldom mentioned in recent scholarship. However, it enjoyed renewed success in the 1950s, just after Reichenbach's death in 1953. In a 1955 paper, Adolf Grünbaum [17], warned from a "widespread error" to consider the "the Lorentz-Fitzgerald contraction hypothesis was an *ad hoc* explanation" as opposed to the Einstein contraction which is a consequence of the relativity of simultaneity [17, 460]. According to Grünbaum this "error is inspired by the numerical equality of the contraction factors of these two kinds of contraction" [17, 460]. Both special relativity and Lorentz theory presupposes the Lorentz contraction: "using light as the standard for effecting the comparison, this hypothesis affirms that in the same system and under the same conditions of measurement, the metrical properties of the arm are different from the ones predicted by classical ether theory" [17, 460]. This has nothing to do with a comparison of the length of the two arms in two different frames. Grünbaum mentions only at the end of the paper that his "treatment of several of the issues is greatly indebted to two outstanding works on the philosophy of relativity by Hans Reichenbach, which are not available in English" [17, 460].

The translation of Reichenbach's book was published in 1958 [53]. A year later, the English translation [43] of Popper's *Logik der Forschung* [42] also appeared in print. Just thereafter, Grünbaum made implicit use of Reichenbach's account against Popper's claim that the Lorentz contraction is *ad hoc* [18]. In response to Popper, Grünbaum [18] drew attention to the 1932 Kennedy-Thorndike experiment, which would disprove Lorentz contraction, if the theory is not supplemented with a clock retardation hypothesis. Grünbaum shows that this "doubly-amended theory" can account for all optical experiments that support special relativity, the Michelson-Morley experiment, the Kennedy-Thorndike experiment, and the Ives-Stillwell experiment. Further modifications can be introduced to account for electromagnetic kind of ether-drift experiments [25]. A few years later, Grünbaum implemented this line of reasoning in his classical monograph, *Philosophical Problems of Space and Time*. The distinction was later adopted in a modified version by Carlo Giannoni [15, 16] and with a different nomenclature by Dennis Dieks [4].

In Reichenbach's view, both Lorentz's and Einstein's theories start with the recognition of a coincidence, that "field geometry" and "matter geometry" always agree, that is that the laws of nature governing field and matter happen to be Lorentz invariant. This non-trivial coincidence requires an explanation. Lorentz's theory, according to Reichenbach, provided the wrong kind of explanation, however, Einstein's theory did not provide any explanation at all. According to the mainstream interpretation [41], it was Minkowski that, so to say, provided the missing explanatory element, i.e. the geometrical structure spacetime [28]. On the contrary, Reichenbach surprisingly adopted a sort of neo-Lorentzian approach [1], claiming that the explanation should have been sought in a future theory of matter. Reichenbach's reaction to Miller's experiments unwittingly shows which is the key difference between these two types of explanations. The "geometrical" explanation is rigid: a positive result of Miller's experiments would have made the entire theory collapse since the geometrical structure of spacetime enters in the formulation of all laws of nature. The "material" explanation is, on the contrary, flexible: the laws governing matter happen to agree with laws governing the fields, but they do not need to do so. The positive result of an ether-drift experiment would have simply implied a readjustment in their reciprocal relationships. Reichenbach regarded this flexibility as an advantage of his axiomatization, whereas it was probably its major shortcomings. Reichenbach's axiomatization seems to miss the the key feature of Einstein-Minkowski approach, which lies in its giving a single explanation for what is otherwise an odd coincidence [27]. The Reichenbach's neo-Lorentzian approach, on the contrary, simply accepts this coincidence as a "brute fact", thus, ultimately, does not provide any explanation at all.

References

[1] Harvey R. Brown. *Physical Relativity. Space-time Structure from a Dynamical Perspective*. Clarendon, 2005.

[2] Rudolf Carnap. Archives of Scientific Philosophy. The Rudolf Carnap Papers.

[3] Leo Corry. *David Hilbert and the Axiomatization of Physics (1898-1918). From* Grundlagen der Geometrie *to* Grundlagen der Physik. Kluwer, 2004.

[4] Dennis Dieks. The 'Reality' of the Lorentz Contraction. *Zeitschrift für allgemeine Wissenschaftstheorie*, 15(2):330–342, 1984.

[5] Jeroen van Dongen. On the Role of the Michelson-Morley Experiment: Einstein in Chicago. *Archive for History of Exact Sciences*, 63(6):655–663, 2009-07.

[6] Arthur Stanley Eddington. *The Mathematical Theory of Relativity*. University Press, 1923.

[7] Arthur Stanley Eddington. *Relativitätstheorie in Mathematischer Behandlung*. Springer, 1925. German translation of [6].

[8] Arthur Stanley Eddington. Ether-drift and the Relativity Theory. *Nature*, 115:870, 1925.

[9] Albert Einstein. The Albert Einstein Archives at the Hebrew University of Jerusalem.

[10] Albert Einstein. Relativitätsprinzip und die aus demselben gezogenen Folgerungen. *Jahrbuch der Radioaktivität und Elektronik*, 4:411–462, 1908. Repr. in [14, Vol. 2, Doc. 47].

[11] Albert Einstein. Die Relativitätstheorie. In Paul Hinneberg, editor, *Die Kultur der Gegenwart*, pages Part 3, Sec. 1, Vol. 3: *Physik*. Ed. by Emil Warburg, 703–713. Teubner, 1915. Repr. in [14, Vol. 4, Doc. 21].

[12] Albert Einstein. Meine Theorie und Millers Versuche. *Vossiche Zeitung*, 19, 1926.

[13] Albert Einstein. Neue Experimente über den Einfluß der Erdbewegung auf die Lichtgeschwindigkeit relativ zur Erde. *Forschungen und Fortschritte*, 3(1, 36):36–37, 1927.

[14] Albert Einstein. *The Collected Papers of Albert Einstein*. Princeton University Press, 1987-.

[15] Carlo Giannoni. Einstein and the Lorentz-Poincaré Theory of Relativity. In Roger C. Buck and Robert S. Cohen, editors, *PSA 1970*, pages 575–589. Reidel, 1971.

[16] Carlo Giannoni. Relativistic Mechanics and Electrodynamics without One-way Velocity Assumptions. *Philosophy of Science*, 45(1):17–46, 1978.

[17] Adolf Grünbaum. Logical and Philosophical Foundations of the Special Theory of Relativity. *American Journal of Physics*, 23(7):450–464, 1955.

[18] Adolf Grünbaum. The Falsifiability of the Lorentz-Fitzgerald Contraction Hypothesis. *The British Journal for the Philosophy of Science*, 10 (37):48–50, 1959.

[19] Klaus Hentschel. Zur Rolle Hans Reichenbachs in den Debatten um die Relativitätstheorie (mit der vollständigen Korrespondenz Reichenbach-Friedrich Adler im Anhang). *Nachrichtenblatt der Deutschen Gesellschaft für Geschichte der Medizin, Naturwissenschaft & Technik*, 3:295–324, 1982.

[20] Klaus Hentschel. *Interpretationen und Fehlinterpretationen der speziellen und der allgemeinen Relativitätstheorie durch Zeitgenossen Albert Einsteins*. Birkhäuser, 1990.

[21] Klaus Hentschel. Einstein's Attitude Towards Experiments. Testing Relativity Theory 1907–1927. *Studies in History and Philosophy of Science*, 23(4):593–624, 1992.

[22] Herbert E. Ives and G. R. Stilwell. An Experimental Study of the Rate of a Moving Atomic Clock. *Journal of the Optical Society of America*, 28(7):215–226, Jul 1938.

[23] Herbert E. Ives and G. R. Stilwell. An Experimental Study of the Rate of a Moving Atomic Clock II. *Journal of the Optical Society of America*, 31(5):369, 1941-05.

[24] Allen Janis. Conventionality of Simultaneity. In Edward N. Zalta, editor, *Stanford Encyclopedia of Philosophy*. Fall edition, 2014.

[25] Michel Janssen. *A Comparison between Lorentz's Ether Theory and Special Relativity in the Light of the Experiments of Trouton and Noble*. PhD thesis, University of Pittsburgh, 1995.

[26] Michel Janssen. Reconsidering a Scientific Revolution. The Case of Einstein Versus Lorentz. *Physics in Perspective*, 4(4):421–446, 2002.

[27] Michel Janssen. COI Stories: Explanation and Evidence in the History of Science. *Perspectives on Science*, 10(4):457–522, 2002.

[28] Michel Janssen. Drawing the Line between Kinematics and Dynamics in Special Relativity. *Studies in History and Philosophy of Science Part B: Studies in History and Philosophy of Modern Physics*, 40(1):26–52, 2009.

[29] Roy J. Kennedy and Edward M. Thorndike. Experimental Establishment of the Relativity of Time. *Physical Review*, 42(3):400–418, Nov 1932.

[30] Roberto Lalli. The Reception of Miller's Ether-drift Experiments in the USA. *Annals of Science*, 69(2):153–214, 2012.

[31] Roberto Lalli. Anti-relativity in Action. The Scientific Activity of Herbert E. Ives between 1937 and 1953. *Historical Studies in the Natural Sciences*, 43(1):41–104, 2013.

[32] Albert A. Michelson. The Relative Motion of the Earth and the Luminiferous Ether. *American Journal of Science*, 22:120–129, 1881.

[33] Albert A. Michelson and Edward W. Morley. On the Relative Motion of the Earth and the Luminiferous Ether. *American Journal of Science*, 34:333–345, 1887.

[34] Dayton C. Miller. Ether-Drift Experiments at Mount Wilson in 1921 and at Cleveland in 1922. *Science*, 55:496, 1922.

[35] Dayton C. Miller. Ether-Drift Experiments at Mount Wilson. *Proceedings of the National Academy of Sciences of the United States of America*, 11(6):306–314, 1925.

[36] Dayton C. Miller. Report on Ether Drift Experiments. *Annual Report of the National Academy of Sciences*, pages 48–49, 1925.

[37] Dayton C. Miller. The Ether-Drift Experiment and the Determination of the Absolute Motion of the Earth. *Review of Modern Physics*, 5: 203–242, 1933.

[38] Edward W. Morley and Dayton C. Miller. On the Theory of Experiments to detect Aberrations of the Second Degree. *Philosophical Magazine*, 9-6th Series(53):669–680, 1905.

[39] Edward W. Morley and Dayton C. Miller. Report of an Experiment to Detect the Fitzgerald-Lorentz Effect. *Proceedings of the American Academy of Arts and Sciences*, 61(12):321–328, 1905.

[40] Edward W. Morley and Dayton C. Miller. Final Report on Ether-Drift Experiments. *Science*, 25(12):525, 1907.

[41] Vesselin Petkov. *Relativity and the Nature of Spacetime*. Springer, 2005.

[42] Karl R. Popper. *Logik der Forschung. Zur Erkenntnistheorie der Modernen Naturwissenschaft*. Springer, 1935.

[43] Karl R. Popper. *The Logic of Scientific Discovery*. Basic Books, 1959.

[44] Hans Reichenbach. Archives of Scientific Philosophy. The Hans Reichenbach Papers.

[45] Hans Reichenbach. Bericht über eine Axiomatik der Einsteinschen Raum-Zeit-Lehre. *Physikalische Zeitschrift*, 22:683–687, 1921.

[46] Hans Reichenbach. La signification philosophique de la théorie de la relativité. *Revue philosophique de la France et de l'Étranger*, 93:5–61, 1922.

[47] Hans Reichenbach. *Axiomatik der relativistischen Raum-Zeit-Lehre.* Braunschweig, 1924. Repr. in [55, Vol. 3, 3–171].

[48] Hans Reichenbach. Über die physikalischen Konsequenzen der relativistischen Axiomatik. *Zeitschrift für Physik*, 34(1):32–48, 1925. Repr. in [55, Vol. 3, 172–183].

[49] Hans Reichenbach. Die Probleme der modernen Physik. *Die Neue Rundschau*, April:414–425, 1926.

[50] Hans Reichenbach. Ist die Relativitätstheorie widerlegt? *Die Umschau*, 30(17):325–328, 1926.

[51] IIans Reichenbach. Metaphysik und Naturwissenschaft. *Symposion*, 1: 158–176, 1926.

[52] Hans Reichenbach. *Philosophie der Raum-Zeit-Lehre.* Walter de Gruyter, 1928. Repr. in [55, Vol. 2].

[53] Hans Reichenbach. *The Philosophy of Space and Time.* Dover Publ., 1958.

[54] Hans Reichenbach. *Axiomatization of the Theory of Relativity.* University of California Press, 1969.

[55] Hans Reichenbach. *Gesammelte Werke in 9 Bänden.* Vieweg, 1977.

[56] Hans Reichenbach. *Selected Writings. 1909–1953.* Reidel, 1978.

[57] Hans Reichenbach. *Defending Einstein. Hans Reichenbach's Writings on Space, Time, and Motion.* Cambridge University Press, 2006.

[58] Thomas Ryckman. *The Reign of Relativity. Philosophy in Physics 1915-1925.* Oxford University Press, 2005.

[59] Robert Rynasiewicz. Weyl vs. Reichenbach on Lichtgeometrie. In A. J. Kox and Jean Eisenstaedt, editors, *The Universe of General Relativity*, pages 137–156. Birkhäuser, 2005.

[60] Moritz Schlick. Die philosophische Bedeutung des Relativitätsprinzips. *Zeitschrift für Philosophie und philosophische Kritik*, 159(2):129–175, 1915.

[61] Moritz Schlick. Die Relativitätstheorie in der Philosophie. In Alexander Wittig, editor, *Verhandlungen der Gesellschaft Deutscher Naturforscher und Ärzte*, pages 58–69. Voge, 1923. Repr. in [63, Vol. 1.5, 529–547].

[62] Moritz Schlick. Erleben, Erkennen, Metaphysik. *Kant-Studien*, 31(2/3): 146– 158, 1926.

[63] Moritz Schlick. *Gesamtausgabe*. Springer, 2006.

[64] Mortitz Schlick. Schlick Nachlass. Noord-Hollands Archief, Haarlem.

[65] Robert S. Shankland. Conversations with Albert Einstein. *American Journal of Physics*, 31(1):47–57, 1963-01.

[66] Robert S. Shankland, Sidney W. McCuskey, Fred C. Leone, and Gustav Kuerti. New Analysis of the Interferometer Observations of Dayton C. Miller. *Reviews of Modern Physics*, 27(2):167–178, 1955-04.

[67] Ludwik Silberstein. D. C. Miller's Recent Experiments, and the Relativity Theory. *Nature*, 115:798–799, May 1925.

[68] Edwin E. Slosson. The Relativity Theory and the Ether Drift. *Science*, 62(Supplement):viii, 1925.

[69] Charles Edward St. John. Observational Basis of General Relativity. *Publications of the Astronomical Society of the Pacific*, 44:277, 1932.

[70] John Stachel. Einstein and Michelson. The Context of Discovery and the Context of Justification. *Astronomische Nachrichten*, 303:47–53, 1982. Repr. in [71, 177–190].

[71] John Stachel. *Einstein from 'B' to 'Z'*. Birkhäuser, 2002.

[72] Loyd S. Swenson. *The Ethereal Aether. A History of the Michelson-Morley-Miller Aether-drift Experiments, 1880-1930*. University of Texas Press, 1972.

[73] Hans Thirring. Prof. Miller's Ether Drift Experiments. *Nature*, 118: 81–82, July 1926.

[74] Rudolf Tomaschek. Über das Verhalten des Lichtes außerirdischer Lichtquellen. *Annalen der Physik*, 378(1-2):105–126, 1924.

[75] Rudolf Tomaschek. Über Versuche zur Auffindung elektrodynamischer Wirkungen der Erdbewegung in großen Höhen II. *Annalen der Physik*, 78(24):743–756, 1925.

[76] Rudolf Tomaschek. Über Versuche zur Auffindung elektrodynamischer Wirkungen der Erdbewegung in großen Höhen II. *Annalen der Physik*, 80(13):509–514, 1926.

[77] J. Weber. Der Michelsonversuch von D. C. Miller auf dem Mount Wilson. *Physikalische Zeitschrift*, 27:5–8, 1926.

[78] Hermann Weyl. Elektrizität und Gravitation. *Physikalische Zeitschrift*, 21(21):649–650, 1920. Repr. in [80, Vol. 2, Doc. 40].

[79] Hermann Weyl. Review of [47]. *Deutsche Literaturzeitung*, 45:2122–2128, 1924.

[80] Hermann Weyl. *Gesammelte Abhandlungen*. Springer, 1968.

[81] Edgar Zilsel. Review of [47]. *Die Naturwissenschaften*, 13:407–409, 1925.

2   Could Minkowski have discovered the cause of gravitation before Einstein?

Vesselin Petkov

Abstract   There are two reasons for asking such an apparently unanswerable question. First, Max Born's recollections of what Minkowski had told him about his research on the physical meaning of the Lorentz transformations and the fact that Minkowski had created the full-blown four-dimensional mathematical formalism of spacetime physics before the end of 1907 (which could have been highly improbable if Minkowski had not been developing his own ideas), both indicate that Minkowski might have arrived at the notion of spacetime independently of Poincaré (who saw it as nothing more than a mathematical space) and at a deeper understanding of the basic ideas of special relativity (which Einstein merely postulated) independently of Einstein. So, had he lived longer, Minkowski might have employed successfully his program of regarding four-dimensional physics as spacetime geometry to gravitation as well. Moreover, Hilbert (Minkowski's closest colleague and friend) had derived the equations of general relativity simultaneously with Einstein. Second, even if Einstein had arrived at what is today called Einstein's general relativity before Minkowski, Minkowski would have certainly reformulated it in terms of his program of geometrizing physics and might have represented gravitation *fully* as the manifestation of the non-Euclidean geometry of spacetime (Einstein regarded the geometrical representation of gravitation as pure mathematics) exactly like he reformulated Einstein's special relativity in terms of spacetime.

1   Introduction

On January 12, 1909, only several months after his Cologne lecture *Space and Time* [1], at the age of 44 Hermann Minkowski untimely left this world. We will never know how physics would have developed had he lived longer. What seems undeniable is that the discovery of the link between gravitation and the non-Euclidean geometry of spacetime might have been quite different

A. S. Stefanov, M. Giovanelli (Eds), *General Relativity 1916 - 2016. Selected peer-reviewed papers presented at the Fourth International Conference on the Nature and Ontology of Spacetime, dedicated to the 100th anniversary of the publication of General Relativity, 30 May - 2 June 2016, Golden Sands, Varna, Bulgaria* (Minkowski Institute Press, Montreal 2017). ISBN 978-1-927763-46-9 (softcover), ISBN 978-1-927763-47-6 (ebook).

from what had actually happened.

On the one hand, Einstein's way of thinking based on conceptual analyses and thought experiments now seems to be perhaps the only way powerful enough to decode the unimaginable nature of gravitation. Indeed in 1907 (most probably in November) Einstein had already been well ahead of Minkowski in terms of deeply thinking of the apparently self-evident manifestations of gravitational phenomena when he made a gigantic step towards the new theory of gravitation [2]:

> I was sitting in a chair in the patent office at Bern when all of a sudden a thought occurred to me: "If a person falls freely he will not feel his own weight." I was startled. This simple thought made a deep impression on me. It impelled me toward a theory of gravitation.

Einstein had been so impressed by this insight that he called it the "happiest thought" of his life [2]. And indeed this is a crucial point – at that time it seemed Einstein had been the only human who realized that no gravitational force acted on a falling body (in fact, as we will see is Section 3 Einstein might have misinterpreted his happiest thought). Then he struggled for eight years to come up with a theory – his general relativity – according to which (as we see it today) gravity is not a force but a manifestation of the curvature of spacetime.

On the other hand, however, Minkowski's three papers on relativity, particularly his Cologne lecture *Space and Time* revealed that in the reformulation of Einstein's special relativity he employed a powerful research strategy (rivaling Einstein's research strategy) – exploring the internal logic of the mathematical formalism of physical theories. That is why, had he lived longer, Minkowski and his closest colleague and friend David Hilbert might have formed an unbeatable team in theoretical physics and might have discovered general relativity (surely under another name) before Einstein. Moreover, contrary to common belief, as Lehmkuhl showed [3], Einstein himself did not believe that general relativity geometrized gravitation: "I do not agree with the idea that the general theory of relativity is geometrizing Physics or the gravitational field" [4].

As there is no way to reconstruct what might have happened in the period 1909-1915 I will outline here what steps had been logically available to Minkowski on the basis of his results. I will imagine two logically possible scenarios. In Section 2 I will describe how Minkowski, while employing his program of geometrizing physics to gravitation, might have realised that gravitational phenomena may be manifestations of a non-Euclidean geometry of spacetime. In Section 3 I will discuss the possibility that it was Einstein who first realized that gravitation can be described in terms of non-Euclidean geometry, but since he regarded the geometrization of gravitation only as a mathematical tool, Minkowski might have reformulated Einstein's general relativity by demonstrating that gravitation is not a physical interaction but just curved-spacetime geometry.

2   First scenario

In order to understand better what Minkowski could have done, had he lived longer, it is important to take explicitly into account two indications of why he appears to have realized independently the equivalence of the times of inertial observers in relative motion (what Einstein postulated and which formed the basis of his special relativity) and that the Lorentz transformations can be regarded as rotations in a four-dimensional world (which was first published by Poincaré but he did not see anything revolutionary in that observation since he believed that physical theories do not necessarily represent anything in the physical world since they are nothing more than *convinient descriptions* of physical phenomena).

These two intications are:

- Max Born's recollections of what Minkowski had told him about his research on the physical meaning of the Lorentz transformations and about his shock when Einstein published his 1905 paper in which he postulated the equivalence of different local times of observers in relative motion.

- What is far more important than Born's recollections is the fully-developed four-dimensional formalism describing an absolute four-dimensional world, which Minkowski reported on December 21, 1907 and the depth of his understanding of the electrodynamics of moving bodies. Such a revolution in both physics and mathematics could not have been possible if Minkowski had not been developing his own ideas but had to first understand Einstein's 1905 paper even better than Einstein in order to invent that formalism to reformulate his theory as a theory of an absolute four-dimensional world. Born's recollections simply confirm what appears to be the most probable history of spacetime physics – that Minkowski independently discovered (i) the equivalence of the times of inertial observers in relative motion, and (ii) the notion of spacetime, but Einstein and Poincaré published their results first.

Here is the historical context of Minkowski's comments reflected in Born's recollections.

By 1905 Minkowski was already internationally recognized as an exceptional mathematical talent – in 1883 he received (with Henry Smith) the French Academy's Grand Prize in Mathematics for his innovative geometric approach to the theory of quadratic forms and in 1896 he published his major work in mathematics *The Geometry of Numbers* [5].

At that time Minkowski was already interested in an unresolved issue at the very core of fundamental physics – at the turn of the nineteenth and twentieth century Maxwell's electrodynamics showed that light is an electromagnetic wave, which seemed to imply that it propagates in a light carrying medium (the luminiferous ether), but its existence had been put into question since the Michelson and Morley interference experiment failed to detect the Earth's motion in that medium. This puzzling result was in full agreement with the experimental impossibility to detect absolute uniform motion with mechanical means captured in Galileo's principle of relativity – absolute uniform motion cannot be detected by mechanical experiments. The Michelson and Morley experiment showed that absolute uniform motion cannot be detected by elecromagnetic experiments either.

Minkowski's documented active involvement with the electrodynamics of moving bodies began in the summer of 1905 when he and his friend David Hilbert co-directed a seminar in Göttingen on the electron theory (dealing with the electrodynamics of moving bodies). Einstein's paper on special relativity was not published at that time; *Annalen der Physik* received the paper on June 30, 1905. Poincaré's longer paper *Sur la dynamique de l'électron* was not published either; it appeared in 1906. Also, "Lorentz's 1904 paper (with a form of the transformations now bearing his name) was not on the syllabus" [6].

Minkowski's student Max Born, who attended the seminar in 1905, wrote [7]:

> We studied papers by Hertz, Fitzcerald, Larmor, Lorentz, Poincaré, and others but also got an inkling of Minkowski's own ideas which were published only two years later.

Born also recalled what specifically Minkowski had said during the seminar (quoted in [8]):

> I remember that Minkowski occasionally alluded to the fact that he was engaged with the Lorentz transformations, and that he was on the track of new interrelationships.

Again Born wrote in his autobiography about what Minkowski had told him after Minkowski's lecture *Space and Time* given on September 21, 1908 [9]:

> He told me later that it came to him as a great shock when Einstein published his paper in which the equivalence of the different local times of observers moving relative to each other were pronounced; for he had reached the same conclusions independently but did not publish them because he wished first to work out the mathematical structure in all its splendour. He never made a priority claim and always gave Einstein his full share in the great discovery.

An additional indication that Minkowski did not just reformulate Einstein's special relativity in terms of spacetime, but that he discovered the spacetime physics[1] by independently realizing (i) the equivalence of the times

---

[1] That Minkowski had indeed been developing his own ideas and independently formulated the physics of spacetime is confirmed by Born's recollections above – the first two show that Minkowski was already discussing his own ideas at the seminar in the summer of 1905. Note that at that time Einstein's 1905 paper was not published; Minkowski asked Einstein to send him the 1905 paper hardly on October 9, 1907 [10]. It appears Minkowski needed two years – from 1905 to 1907 – to develop the full mathematical formalism of the four-dimensional physics of spacetime introduced by him (published in 1908 as a 59-page treatise [11]).

of inertial observers in relative motion, and (ii) the meaning of the Lorentz transformations (by successfully decoding the profound physical message hidden in the failed experiments to detect absolute uniform motion) is the fact that Minkowski *explained* what Einstein merely *postulated*. Einstein postulated:

- The equivalence of the time of a "stationary" observer and the different time of a moving observer (formally introduced as an auxiliary mathematical notion by Lorentz).

- The experimental impossibility to detect absolute motion (captured in the relativity postulate).

- That the speed of light is the same in all inertial reference frames.

Minkowski explained (see Minkowski's paper [1] and also [13] and in this section):

- The equivalence of the times of inertial observers in relative motion – *why* such observers have different times.

- The relativity postulate – *why* absolute motion (with constant velocity) cannot be detected or its modern formulation – *why* the laws of physics are the same in all inertial reference frames.

- *Why* the speed of light is the same in all inertial reference frames.

It seems it took some time for Einstein to unterstand Minkowski's spacetime physics as implied by Sommerfeld's recollection of what Einstein said on one occasion which reveals Einstein's initial rather hostile attitude towards Minkowski's work: "Since the mathematicians have invaded the relativity theory, I do not understand it myself any more" [14]. Despite his initial negative reaction towards Minkowski's four-dimensional physics Einstein relatively quickly realized that his revolutionary theory of gravitation would be impossible without the revolutionary spacetime physics discovered by Minkowski. At the beginning of his 1916 paper on general relativity Einstein wrote: "The generalization of the theory of relativity has been facilitated considerably by Minkowski, a mathematician who was the first one to recognize the formal equivalence of space coordinates and the time coordinate, and utilized this in the construction of the theory"[2] [15].

To understand fully what logical options would have been realistically available to Minkowski in 1909, one has to realize that Minkowski regarded the unification of space and time into *die Welt* – a four-dimensional world – as real. This is important not only to understand what Minkowski could have done had he lived longer, but because the issue of the reality of spacetime (Minkowski's four-dimensional world) constitutes an unprecedented situation in fundamental physics. It seems many physicists, including relativists, simply refuse to see the double experimental proof of the reality of spacetime. The first experimental proof is the set of all experiments (including the Michelson and Morley experiment) that failed to detect absolute uniform

---

[2]This quote is hardly from the new 1997 translation [15]. Quite strangely, the first page of the paper containing the recognition of Minkowski's work had been omitted in the first English translation [18].

motion and that gave rise to the relativity postulate. It is these experiments whose hidden profound message was successfully decoded by Minkowski – absolute (uniform) motion cannot be detected because such a thing does not exist in Nature; absolute motion presupposes absolute (i.e. single) space, but those experiments imply that observers in relative motion have different times and spaces, which in turn implies that the world is a four-dimensional world.

On September 21, 1908 Minkowski explained how he decoded the profound message hidden in the failed experiments to discover absolute motion in his famous lecture *Space and Time* and announced the revolutionary view of space and time, which he deduced from those experiments [1, p.111]:

> The views of space and time which I want to present to you arose from the domain of experimental physics, and therein lies their strength. Their tendency is radical. From now onwards space by itself and time by itself will recede completely to become mere shadows and only a type of union of the two will still stand independently on its own.

Here is Minkowski's most general proof that the world is four-dimensional. To explain the experiment of Michelson and Morley, which failed to detect the Earth's absolute motion, Lorentz suggested that observers on Earth can formally use a time that is different from the true time of an observer at absolute rest. Einstein postulated that the times of different inertial observers in relative motion are equally good, that is, each observer has his own time, and that for Einstein meant that time is relative.

Minkowski demonstrated that as observers in relative motion have different equally real times, they inescapably have *different spaces* as well, because space is defined as a set of simultaneous events, and different times imply different sets of simultaneous events, i.e., different spaces[3] (or simply – different times imply different spaces because space is perpendicular to time) [1, p. 114]:

> "Hereafter we would then have in the world no more *the* space, but an infinite number of spaces analogously as there is an infinite number of planes in three-dimensional space. Three-dimensional geometry becomes a chapter in four-dimensional physics. You see why I said at the beginning that space and time will recede completely to become mere shadows and only a world in itself will exist."

Therefore the experimental failure to detect absolute motion has indeed a profound physical meaning – that there exists not a single (and therefore absolute) space, but many spaces (and many times). As many spaces are

---

[3] Minkowski specifically tried to explain why "the concept of space was shaken neither by Einstein nor by Lorentz" [1, p. 117] which prevented them from discovering the truly revolutionary spacetime physics.

possible in a four-dimensional world, Minkowski's irrefutable proof that the world is four-dimensional becomes self-evident:[4]

*If the real world were three-dimensional, there would exist a **single** space, i.e. a single class of simultaneous events (and therefore a single time), which would mean that simultaneity and time would be absolute in contradiction with both the theory of relativity and, most importantly, with the experiments which failed to detect absolute motion.*

The second experimental proof of the reality of spacetime are all experiments that confirmed the kinematic relativistic effects. How these experiments would be *impossible* if the world were *not* four-dimensional (i.e., if spacetime were just a mathematical space) is immediately seen in Minkowski's own explanation of length contraction (which is the accepted explanation) – as length contraction (along with time dilation) is a specific manifestation of relativity of simultaneity, an assumption that reality is *not* a four-dimensional world directly leads (as in the above paragraph) to absolute simultaneity (and to the impossibility of length contraction [16]) in contradiction with relativity and the experiments that confirmed length contraction; one of the experimental tests of length contraction (along with time dilation) is the muon experiment – "in the muon's reference frame, we reconcile the theoretical and experimental results by use of the length contraction effect, and the experiment serves as a verification of this effect" [17].

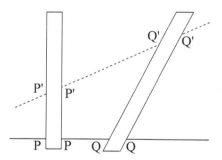

The right half of Figure 1 of Minkowski's paper *Space and Time*

To see exactly how length contraction would be impossible if reality were a three-dimensional world, consider Minkowski's explanation whose essence is that length contraction of a body is a manifestation of the reality of the body's worldtube. Minkowski considered two bodies in uniform relative motion represented by their worldtubes in the figure above (see Figure 1 of Minkowski's paper [1]). Consider only the body represented by the vertical worldtube to understand why the worldtube of a body must be real in order that length contraction be possible. The three-dimensional cross-section PP, resulting from the intersection of the body's worldtube and the space (repre-

---

[4]Minkowski did not bother to state this proof (that if the world were three-dimensional, none of the experiments captured in the relativity postulate, including the Michelson and Morley experiment, would be possible) explicitly; as a mathematicion he believed it was indeed self-evident.

sented by the horizontal line in the figure) of an observer at rest with respect to the body, is the body's proper length. The three-dimensional cross-section $P'P'$, resulting from the intersection of the body's worldtube and the space (represented by the inclined dashed line) of an observer at rest with respect to the second body (represented by the inclined worldtube), is the relativistically contracted length of the body measured by that observer (one should always keep in mind that the cross-section $P'P'$ only looks longer than PP because a fact of the pseudo-Euclidean geometry of spacetime is represented on the Euclidean surface of the page).

Now assume that the worldtube of the body did not exist as a four-dimensional object and were merely an abstract geometrical construction. Then, what would exist would be a single three-dimensional body, represented by the proper cross-section PP, and both observers would measure the *same* three-dimensional body PP of the *same* length. Therefore, not only would length contraction be *impossible*, but relativity of simultaneity would be also impossible since a spatially extended three-dimensional object is defined in terms of *simultaneity* – as all parts of a body taken *simultaneously* at a given moment.[5] Because both observers in relative motion would measure the same three-dimensional body (represented by the cross-section PP) they would share the *same* class of simultaneous events (therefore simultaneity would turn out to be absolute) in contradiction with relativity and with the experiments that confirmed the specific manifestations of relativity of simultaneity – length contraction and time dilation.

All experiments that confirmed time dilation and the twin paradox effect are also impossible in a three-dimensional world [19]. For example, it is an experimental fact, used every second by the GPS, that observers in relative motion have different times, which is impossible in a three-dimensional world [19].

I think the unprecedented situation in fundamental physics – ignoring the fact that the relativistic experiments and the theory of relativity itself are impossible in a three-dimensional world[6] – should be faced and addressed because this situation prevents a proper understanding of the physical meaning of general relativity as revealing that gravitational phenomena are nothing more than a manifestation of the curvature of spacetime; such a deep understanding of the nature of gravity may have important implications for the research on quantum gravity and on gravitational waves.

After Minkowski explained in his lecture *Space and Time* that the true reality is a four-dimensional world in which all ordinarily perceived three-

---

[5] The fact that an extended three-dimensional body is defined in terms of simultaneity confirms Minkowski's interpretation of the cross-sections PP and $P'P'$ as two three-dimensional bodies – while measuring the *same* body, the two observers measure *two* three-dimensional bodies represented by the two cross-sections. This relativistic situation only looks paradoxical at first sight because what is meant by "the same body" is the body's worldtube; the cross-sections PP and $P'P'$ represent the two three-dimensional bodies measured by the two observers.

[6] It appears to be a real problem in physics that some physicists regard issues such as the reality of spacetime as belonging to philosophy, which is *physics at its worst* - the issue of the dimensionality of the world is pure physics.

dimensional particles are a forever given web of worldlines, he outlined his ground-breaking idea of regarding physics as spacetime geometry [1, p. 112]:

> The whole world presents itself as resolved into such worldlines, and I want to say in advance, that in my understanding the laws of physics can find their most complete expression as interrelations between these worldlines.

Then he started to implement his program by explaining that inertial motion is represented by a timelike *straight* worldline, after which he pointed out that [1, p. 115]:

> With appropriate setting of space and time the substance existing at any worldpoint can always be regarded as being at rest.

In this way he explained not only *why* the times of inertial observers are equivalent (their times can be chosen along their timelike worldlines and all straight timelike worldlines in spacetime are equivalent) but also the physical meaning of the relativity principle – the physical laws are the same for all inertial observers (inertial reference frames), i.e. all physical phenomena look exactly the same for all inertial observers, because every observer describes them in his own space (in which he is at rest) and uses his own time. For example the speed of light is the same for all observers because each observer measures it in its own space using his own time.

Then Minkowski explained that accelerated motion is represented by a *curved* or, more precisely, *deformed* worldline and noticed that "Especially the concept of *acceleration* acquires a sharply prominent character."

As Minkowski knew that a particle moving by inertia offers no resistance to its motion with constant velocity (which explains why inertial motion cannot be detected experimentally as Galileo first demonstrated), whereas the accelerated motion of a particle can be discovered experimentally since the particle *resists* its acceleration, he might have very probably linked the sharp physical distinction between inertial (non-resistant) and accelerated (resistant) motion with the sharp geometrical distinction between inertial and accelerated motion represented by straight and deformed (curved) worldlines, respectively.

The realization that an accelerated particle (which resists its acceleration) is a deformed worldtube in spacetime would have allowed Minkowski to notice two virtually obvious implications of this spacetime fact [19]:

- The acceleration of a particle is absolute not because it accelerates with respect to some absolute space, but because its worldtube is deformed, which is an absolute geometrical and physical fact.

- The resistance a particle offers to its acceleration (i.e. its inertia) originates from a four-dimensional stress in its *deformed* worldtube.[7] That is, the

---

[7]Note that the worldtube, and therefore spacetime itself, must be real for this to be possible. The very correspondence between the sharp physical and geometrical distinction of inertial and accelerated motion strongly (and independently) implies that spacetime is real.

inertial force with which the particle resists its acceleration turns out to be a static restoring force arising in the *deformed* worldtube of the accelerated particle. I guess Minkowski might have been particularly thrilled by this implication of his program to regard physics as spacetime geometry because inertia happens to be another manifestation of the fact that reality is a four-dimensional world.

To demonstrates the enormous potential of Minkowski's program of geometrizing physics let us assume that Minkowski had read Galileo's works, particularly Galileo's analysis demonstrating that heavy and light bodies fall at the *same* rate [20]. In this analysis Galileo virtually came to the conclusion that a falling body does not resist its fall [20]:

> But if you tie the hemp to the stone and allow them to fall freely from some height, do you believe that the hemp will press down upon the stone and thus accelerate its motion or do you think the motion will be retarded by a partial upward pressure? One always feels the pressure upon his shoulders when he prevents the motion of a load resting upon him; but if one descends just as rapidly as the load would fall how can it gravitate or press upon him? Do you not see that this would be the same as trying to strike a man with a lance when he is running away from you with a speed which is equal to, or even greater, than that with which you are following him? You must therefore conclude that, during free and natural fall, the small stone does not press upon the larger and consequently does not increase its weight as it does when at rest.

Then the path to the idea that gravitational phenomena are manifestations of the curvature of spacetime would have been open to Minkowski – the experimental fact that a falling particle accelerates (which means that its worldtube is curved), but offers no resistance to its acceleration (which means that its worldtube is not deformed) can be explained only if the worldtube of a falling particle is *both curved and not deformed*, which is impossible in the flat Minkowski spacetime where a curved worldtube is always deformed. Such a worldtube can exist only in a non-Euclidean spacetime whose geodesics are naturally curved due to the spacetime curvature, but are not deformed.

As for Minkowski spacetime (*die Welt*) was real, then it would not have been difficult for him (as a mathematician who listens to what the mathematical formalism tells him and is not affected by the appearance that gravitation is a physical interaction) to realize that gravitational phenomena are fully explained as manifestations of the non-Euclidean geometry of spacetime with no need to assume the existence of gravitational interaction. Indeed, particles fall toward the Earth's surface and planets orbit the Sun not due to a gravitational force or interaction, but because they move by inertia (nonresistantly); expressed in correct spacetime language, the falling particles and planets are geodesic worldlines (or rather worldtubes) in spacetime.

Minkowski would have easily explained the force acting on a particle on the Earth's surface, i.e. the particle's weight. The worldtube of a particle

42

falling toward the ground is geodesic, which, in ordinary language, means that the particle moves by inertia (non-resistantly). When the particle lands on the ground it is prevented from moving by inertia and it resists the change of its inertial motion by exerting an inertial force on the ground. Like in flat spacetime the inertial force originates from the *deformed* worldtube of the particle which is at rest on the ground.[8] So the weight of the particle that has been traditionally called gravitational force would turn out to be inertial force, which naturally explains the observed equivalence of inertial and gravitational forces. While the particle is on the ground its worldtube is deformed (due to the curvature of spacetime), which means that the particle is being constantly subjected to a curved-spacetime acceleration (keep in mind that acceleration means deformed worldtube!); the particle resists its acceleration through the inertial force and the measure of the resistance the particle offers to its acceleration is its inertial mass, which traditionally has been called (passive) gravitational mass. This fact naturally explains the equivalence between a particle's inertial and gravitational masses, which turned out to be the same thing.

In this way, Minkowski would have again *explained* one more set of experimental facts which Einstein merely *postulated* – Einstein "explained" these experimental facts by his equivalence postulate. So Minkowski would have explained Einstein's equivalence postulate exactly like he explained Einstein's relativity postulate.

3   Second scenario

Now imagine that after his lecture *Space and Time* Minkowski found a very challenging mathematical problem and did not compete with Einstein for the creation of the modern theory of gravitation. But when Einstein linked gravitation with the geometry of spacetime Minkowski regretted his change of research interests and started to study intensely general relativity and its implications.

As a mathematician Minkowski would be greatly impressed by the genius of his former student Einstein for linking gravitation with the geometry of spacetime and by the elegent mathematical formalism developed by Einstein with the help of another former student (and a friend of Einstein) – Marcel Grossmann. At the same time Minkowski would be appalled by Einstein's inability to trust the mathematical formalism of his general relativity and to try to smuggle into the theory the apparently self-evident notions of gravitational interaction and gravitational energy.

Minkowski would see Einstein's general relativity as a triumph of his program of geometrizing physics and would reformulate, or rather properly interpret, general relativity by pointing out that:

---

[8]Note again that Minkowski would have explained this fact only because he regarded spacetime as real – a fact deduced from all failed *experiments* designed to detect absolute uniform motion.

- The new theory of gravitation demonstrates that gravitational phenomena are in fact nothing more than manifestations of the non-Euclidean geometry of spacetime.

- General relativity *itself* demonstrates that gravitational phenomena are *fully* explained by the non-Euclidean geometry of spacetime and are not caused by gravitational interaction – particles falling toward the Earth and planets orbiting the Sun *all move by inertia* and inertia by its very nature presupposes *no interaction*. In the correct spacetime language the falling particles' worldlines and the planets' worldlines are geodesics which represent inertial (i.e. *non-resistant*[9]) motion.

- There is no gravitational field and no gravitational force in Nature – the weight of a particle on the Earth's surface which has always, before the advent of general relativity, been regarded as a gravitational force (caused by the Earth's gravitational field) is, according to a proper understanding of the mathematical formalism of general relativity (and as Minkowski would have found as we saw in the first scenario), *inertial* force.[10]

- As a mathematician Minkowski would point out that the mathematical formalism of general relativity provides additional proof that gravitational phenomena are not caused by gravitational *interaction* – the mathematical formalism of general relativity *itself* refuses to yield a proper (tensorial) expression for gravitational energy and momentum, which demonstrates that these are not present in the physical world. Moreover, the fact that "in relativity there is no such thing as the force of gravity" [21] implies that there is no gravitational energy either since such energy is defined as the work done by gravitational forces. Whether or not gravitational energy is regarded as local does not affect the very definition of energy.

- Minkowski's approach to understanding gravitational phenomena would help him identify the major open question in gravitational physics – *how*

---

[9]It is an experimental fact that particles falling toward the Earth's surface *do not resist* their fall – a falling accelerometer, for example, reads zero resistance (i.e. zero acceleration; the observed *apparent* acceleration of the accelerometer is caused by the spacetime curvature caused by the Earth). The experimental fact that particles do not resist their fall (i.e. their apparent acceleration) means that they move by inertia and therefore no gravitational force is causing their fall. It should be emphasized that a gravitational force would be required to accelerate particles downwards *if and only if* the particles *resisted* their acceleration, because *only then* a gravitational force would be needed to *overcome* that resistance.

[10]Einstein believed (as the quote in the Introduction reveals) that the geometrization of gravitation is nothing more than a mathematical representation of real gravitational interaction with real gravitational force and energy. Therefore, it seems Einstein had misinterpreted his "happiest thought" – he might have believed that the gravitational force acting on a particle, causing its fall, is somehow compensated by the inertial force with which the particle resists its downward acceleration (in line with his equivalence principle). However, that would not explain his "happiest thought" that a falling person "will not feel his own weight," because if there were a gravitational force acting on the person, *his fall would not be non-resistant* – his body will *resist* the gravitational force which accelerates it downwards (exactly like a particle accelerated by a force in open space resists its acceleration); the very physical meaning of the inertial force is that it is a *resistance* force, with which a particle resists its acceleration.

matter changes the geometry of spacetime.

4   Instead of Conclusion

Gravitation as a separate agency becomes unnecessary
Arthur S. Eddington [22]

*An electromagnetic field is a "thing;" gravitational field
is not, Einstein's theory having shown that it is nothing
more than the manifestation of the metric*
Arthur S. Eddington [23]

Despite that taken at face value general relativity fully explains gravitational phenomena without assuming that there exists gravitational interaction, there have been continuing attempts (initiated by Einstein) to smuggle the concept of gravitational interaction into the framework and mathematical formalism of general relativity.

Despite the arguments Minkowski would have pointed out (listed above), the prevailing view among relativists is that there exists indirect astrophysical evidence for the existence of gravitational energy – coming from the interpretation of the decrease of the orbital period of the binary pulsar system PSR 1913+16 discovered by Hulse and Taylor in 1974 [24] (and other such systems discovered after that), which is believed to be caused by the loss of energy due to gravitational waves emitted by the system (which carry away gravitational energy).

This interpretation that gravitational waves carry gravitational energy should be carefully scrutinized (especially after the recent detection of gravitational waves) by taking into account the arguments against the existence of gravitational energy and momentum and especially the fact that there does not exist a rigorous (analytic, proper general-relativistic) solution for the two body problem in general relativity. I think the present interpretation of the decrease of the orbital period of binary systems contradicts general relativity, particularly the geodesic hypothesis (geodesics represent inertial motion) and the experimental evidence (falling particles do not resist their fall) which confirmed it, because by the geodesic hypothesis the neutron stars, whose worldlines had been regarded as exact geodesics (since the stars had been modelled dynamically as a pair of orbiting *point* masses by Hulse and Taylor), *move by inertia without losing energy since the very essence of inertial motion is motion with no loss of energy.* For this reason no energy can be carried away by the gravitational waves emitted by the binary pulsar system. Let me stress it as strongly as possible: the geodesic hypothesis (confirmed by experiment) and the assertion that bodies, whose worldlines are geodesics, emit gravitational energy (carried away by gravitational waves), cannot be both correct.

In fact, it is the very assumption that the binary system emits gravitational waves which contradicts general relativity in the first place, because motion by inertia does not generate gravitational waves in general relativity. The inspiralling neutron stars in the binary system were modelled as *point* masses and therefore their worldlines are exact geodesics, which means that the stars move by inertia and no emission of gravitational radiation is involved; if the stars were modelled as extended bodies, then and only then they would be subject to tidal effects and energy would be involved, but that energy would be negligibly small (see next paragraph) and would not be gravitational (see the explanation of the origin and nature of energy in the sticky bead argument below). So, the assertion that the inspiralling neutron stars in the binary system PSR 1913+16 generate gravitational waves is incorrect because it contradicts general relativity.

Gravitational waves are emitted only when the stars' timelike worldlines are not geodesic,[11] that is, when the stars are subject to an absolute (curved-spacetime) acceleration (associated with the absolute feature that a worldline is not geodesic), not a relative (apparent) acceleration between the stars caused by the geodesic deviation of their worldlines. For example, in general relativity the stars are subject to an absolute acceleration when they collide (because their worldlines are no longer geodesic); therefore gravitational waves – carrying no gravitational energy-momentum – are emitted only when the stars of a binary system collide and merge into one, that is, "Inspiral gravitational waves are generated during the end-of-life stage of binary systems where the two objects merge into one" [25].

Let me repeat it: when the stars follow their orbits in the binary system, they do not emit gravitational waves since they move by inertia according to general relativity (their worldlines are geodesic and no absolute acceleration is involved); even if the stars were modelled as extended bodies, the worldlines of the stars' constituents would not be geodesic (but slightly deviated from the geodesic shape) which will cause tidal friction in the stars, but the gravitational waves generated by the very small absolute accelerations of the stars' constituents will be negligibly weak compared to the gravitational waves believed to be emitted from the spiralling stars of the binary system (that belief arises from using not the correct general-relativistic notion of acceleration ($a^\mu = d^2x^\mu/d\tau^2 + \Gamma^\mu_{\alpha\beta}(dx^\alpha/d\tau)(dx^\beta/d\tau)$), but the Newtonian one).

The famous sticky bead argument has been regarded as a decisive argu-

---

[11] The original prediction of gravitational wave emission, obtained by Einstein (*Berlin. Sitzungsberichte*, 1916, p. 688; 1918, p. 154), correctly identified the source of such waves – a spinning rod, or any rotating material bound together by cohesive force. None of the particles of such rotating material (except the centre of rotation) are geodesic worldlines in spacetime and, naturally, such particles will emit gravitational waves. This is not the case with double stars; as the stars are modelled as point masses, their worldliness are exact geodesics (which means that the stars are regarded as moving by inertia) and no gravitational waves are emitted. If the stars are regarded as extended bodies their worldtubes will still be geodesic, but their motion will not be entirely non-resistant, because of the tidal friction within the stars (caused by the fact that the worldlines of the stars' constituents are not congruent due to geodesic deviation).

ment in the debate on whether or not gravitational waves transmit gravitational energy because it has been perceived to demonstrate that gravitational waves do carry gravitational energy which was converted through friction into heat energy [26]:

> The thought experiment was first described by Feynman (under the pseudonym "Mr. Smith") in 1957, at a conference at Chapel Hill, North Carolina. His insight was that a passing gravitational wave should, in principle, cause a bead which is free to slide along a stick to move back and forth, when the stick is held transversely to the wave's direction of propagation. The wave generates tidal forces about the midpoint of the stick. These produce alternating, longitudinal tensile and compressive stresses in the material of the stick; but the bead, being free to slide, moves along the stick in response to the tidal forces. If contact between the bead and stick is 'sticky,' then heating of both parts will occur due to friction. This heating, said Feynman, showed that the wave did indeed impart energy to the bead and rod system, so it must indeed transport energy.

However, a careful examination of this argument reveals that kinetic, not gravitational, energy is converted into heat because a gravitational wave changes the shape of the geodesic worldline of the bead (and of the stick) and the stick prevents the bead from following its changed geodesic worldline, i.e., prevents the bead from moving by inertia; as a result the bead resists and exerts an *inertial* force on the stick (exactly like when a particle away from gravitating masses moving by inertia is prevented from its inertial motion, it exerts an inertial force on the obstacle and the kinetic energy of the particle is converted into heat).

It appears more adequate if one talks about *inertial*, not kinetic, energy, because what is converted into heat (as in the sticky bead argument) is the energy corresponding to the work done by the inertial force (and it turns out that that energy, originating from the inertial force, is equal to the kinetic energy [27]). The need to talk about the adequate inertial, not kinetic, energy is clearly seen in the explanation of the sticky bead argument above – initially (before the arrival of the gravitational wave) the bead is at rest and does not possess kinetic energy; when the gravitational wave arrives, the bead starts to move but by inertia (non-resistantly) since the shape of its geodesic worldline is changed by the wave into another geodesic worldline (which means that the bead goes from one inertial state – rest – into another inertial state, i.e., without any transfer of energy from the gravitational wave; transferring energy to the bead would occur if and only if the gravitational wave changed the state of the bead from inertial to non-inertial), and when the stick tries to prevent the bead from moving by inertia, the bead resists and exerts an inertial force on the stick (that is why, what converts into heat through friction is inertial energy).

Finally, it is a fact in the rigorous structure of general relativity that gravitational waves do not carry gravitational energy,[12] which, however, had been inexplicably ignored, despite that Eddington explained it clearly in his comprehensive treatise on the mathematical foundations of general relativity *The Mathematical Theory of Relativity* [23, p. 260]: "The gravitational waves constitute a genuine disturbance of space-time, but their energy, represented by the pseudo-tensor t^{ν}_{μ}, is regarded as an analytical fiction" (it cannot be regarded as an energy of any kind for the well-known reason that "It is not a tensor-density and it can be made to vanish at any point by suitably choosing the coordinates; we do not associate it with any absolute feature of world-structure," *ibid*, p. 136).

References

[1] H. Minkowski, *Space and Time: Minkowski's Papers on Relativity.* Translated by Fritz Lewertoff and Vesselin Petkov. Edited by V. Petkov (Minkowski Institute Press, Montreal 2012)

[2] Quoted from: A. Pais, *Subtle Is the Lord: The Science and the Life of Albert Einstein* (Oxford University Press, Oxford 2005) p. 179

[3] D. Lehmkuhl, Why Einstein did not believe that General Relativity geometrizes gravity. *Studies in History and Philosophy of Physics*, Volume 46, May 2014, pp. 316-326

[4] A letter from Einstein to Lincoln Barnett from June 19, 1948; quoted in [3].

[5] H. Minkowski, *Geometrie der Zahlen* (Teubner, Leipzig 1896)

[6] S. Walter, Minkowski, Mathematicians, and the Mathematical Theory of Relativity, in H. Goenner, J. Renn, J. Ritter, T. Sauer (eds.), *The Expanding Worlds of General Relativity*, Einstein Studies, volume 7, (Birkhäuser, Basel 1999) pp. 45-86, p. 46

[7] M. Born, *Physics in My Generation* 2nd ed. (Springer-Verlag, New York 1969) p. 101

[8] T. Damour, "What is missing from Minkowski's "Raum und Zeit" lecture", *Annalen der Physik* **17** No. 9-10 (2008), pp. 619-630, p. 626

---

[12] An immediate and misleading reaction "A wave that carries no energy?!" should be resisted, because it is from the old times of three-dimensional thinking – assuming that a wave really travels in the external world. There is no such thing as a propagating wave in spacetime – what is there is a spacetime region whose "wavelike" geometry is interpreted in three-dimensional language as a wave which propagates in space (exactly like a timelike worldline is interpreted in three-dimensional language as a particle which moves in space); also, keep in mind that there is no such thing as space in the external world, because spacetime is not divided into a space and a time.

[9] M. Born, *My Life: Recollections of a Nobel Laureate* (Scribner, New York 1978) p. 131

[10] Postcard: Minkowski to Einstein, October 9, 1907, in: M.J. Klein, A. J. Kox, and R. Schulmann (eds) *The Collected Papers of Albert Einstein, Volume 5: The Swiss Years: Correspondence, 1902-1914* (Princeton University Press, Princeton 1995), p. 62.

[11] H. Minkowski, Die Grundgleichungen für die elektromagnetischen Vorgänge in bewegten Körpern, Nachrichten der K. Gesellschaft der Wissenschaften zu Göttingen. *Mathematisch-physikalische Klasse* (1908) S. 53-111; reprinted in H. Minkowski, *Zwei Abhandlungen über die Grundgleichungen der Elektrodynamik, mit einem Einführungswort von Otto Blumenthal* (Teubner, Leipzig 1910) S. 5-57, and in *Gesammelte Abhandlungen von Hermann Minkowski*, ed. by D. Hilbert, 2 vols. (Teubner, Leipzig 1911), vol. 2, pp. 352-404.

[12] H. Poincaré, Sur la dynamique de l'électron, *Rendiconti del Circolo matematico Rendiconti del Circolo di Palermo* **21** (1906) pp. 129-176

[13] V. Petkov, Physics as Spacetime Geometry. In: A. Ashtekar, V. Petkov (eds), *Springer Handbook of Spacetime* (Springer, Heidelberg 2014), pp. 141-163

[14] A. Sommerfeld, To Albert Einstein's Seventieth Birthday. In: *Albert Einstein: Philosopher-Scientist*. P. A. Schilpp, ed., 3rd ed. (Open Court, Illinois 1969) pp. 99-105, p. 102

[15] *The Collected Papers of Albert Einstein, Volume 6: The Berlin Years: Writings, 1914-1917* (Princeton University Press, Princeton 1997), p. 146

[16] V. Petkov, On the Reality of Minkowski Space, *Foundations of Physics* **37** (2007) 1499–1502

[17] G.F.R. Ellis and R.M. Williams, *Flat and Curved Space-Times* (Oxford University Press, Oxford 1988) p. 104

[18] H. A. Lorentz et al., *The Principle of Relativity*, translated by W. Perrett and G. B. Jeffery (Methuen 1923; Dover repr., 1952)

[19] V. Petkov, *Relativity and the Nature of Spacetime*, 2nd ed. (Springer, Heidelberg 2009) Chapter 5

[20] Galileo, *Dialogues Concerning Two Sciences*. In: S. Hawking (ed.), *On The Shoulders Of Giants*, (Running Press, Philadelphia 2002) pp. 399-626, p. 447

[21] J. L. Synge, *Relativity: the general theory*. (Nord-Holand, Amsterdam 1960) p. 109

[22] A.S. Eddington, The Relativity of Time, *Nature* **106**, 802-804 (17 February 1921); reprinted in: A. S. Eddington, *The Theory of Relativity and its Influence on Scientific Thought: Selected Works on the Implications of Relativity* (Minkowski Institute Press, Montreal 2015) pp. 27-30, p. 30

[23] A.S. Eddington, *The Mathematical Theory of Relativity*, Minkowski (Institute Press, Montreal 2016) p. 233

[24] R.A. Hulse, J.H. Taylor, Discovery of a pulsar in a binary system, *Astrophys. J.* **195** (1975) L51–L53

[25] Introduction to LIGO and Gravitational Waves: Inspiral Gravitational Waves http://www.ligo.org/science/GW-Inspiral.php

[26] Sticky bead argument, https://en.wikipedia.org/wiki/Sticky_bead_argument

[27] V. Petkov, On Inertial Forces, Inertial Energy and the Origin of Inertia, published as Appendix B in [28]

[28] V. Petkov, *Inertia and Gravitation: From Aristotle's Natural Motion to Geodesic Worldlines in Curved Spacetime* (Minkowski Institute Press, Montreal 2012).

3 INCONGRUENT COUNTERPARTS AND THE NATURE OF SPACE

ANTONELLA FOLIGNO

Abstract   The paper provides a novel reconstruction of the infamous Incongruent Counterparts argument which traces back to Kant. In particular, it argues that the argument allegedly supports two different conclusions, namely a negative one against the relationalist conception of space, and a positive one in favour of the alternative substantivalist conception. After that, the paper develops two possible responses on behalf of the relationalist that undermine both the previous conclusions. It concludes by assessing whether a purely geometrical argument could ever suffice to vindicate a particular metaphysics of space over another.

1   Introduction

The debate over the nature of space is surely one of the most intense and widely had debates in the study of metaphysics and the philosophy of science. Traditionally there are two leading schools of thought. On the one hand we have substantivalism, which can be summarized[1] as the thesis according to which space is treated as a particular substance that does not depend upon the objects which are located in it. On the other hand we have relationalism, according to which space is nothing but a kind of collection of spatial relations between more fundamental physical objects, and is thus altogether dependent on them for its existence.[2]

The most forceful argument on the side of the substantivalist has been, from Newton's famous 'rotating bucket' argument onwards, that we still lack a purely relational theory of motion.[3] There is however another line of argument tracing back to Kant, which is purely geometrical in nature (thus

---

[1] It is not the aim of the paper to spell out the metaphysical theses about the nature of space more rigorously. The interested reader can start from Earman(1989).

[2] Traditionally these views are linked to the name of Newton and Clarke on the one hand, Leibniz on the other.

[3] But see Barbour and Bertotti (1982) and Barbour (1989).

A. S. Stefanov, M. Giovanelli (Eds), *General Relativity 1916 - 2016. Selected peer-reviewed papers presented at the Fourth International Conference on the Nature and Ontology of Spacetime, dedicated to the 100th anniversary of the publication of General Relativity, 30 May - 2 June 2016, Golden Sands, Varna, Bulgaria* (Minkowski Institute Press, Montreal 2017). ISBN 978-1-927763-46-9 (softcover), ISBN 978-1-927763-47-6 (ebook).

avoiding any reference to a particular kinematics or dynamics), and which alleges to show the inadequacy of the relationalist account of space.

It is this line of argument that my paper focuses on. The outline of my paper is as follows. In Section 2 I will present my own reconstruction of Kant's geometrical argument and in Section 3 I will provide some possible responses on behalf of the relationalist.

2   Incongruent Counterparts: a Geometrical Argument for the Reality of Space

In his work *On the First Ground of the Distinction of Regions of Space (Von demerstenGrund des Unterchiedes der GegendenimRaum,* 1768) Kant draws our attention to particular material bodies which display a striking feature, namely that they resemble each other almost perfectly, yet they cannot be made to be congruent by any means of a simple rigid motion. They are a mirror image of one another. These material objects are called "incongruent counterparts" and it is the existence of these very objects which threatens the relationalist's metaphysics. In this section I will provide my own reconstruction – with no intention of historical accuracy – of the argument pertaining to incongruent counterparts.[4]

Let us first provide a definition of incongruent counterparts that is interesting enough to fuel the following analysis.

Two things (x, y) are said to be incongruent counterparts iff: (i) they cannot be made to be congruent by means of any continuous rigid motion (e.g. by translation or spatial rotation or any suitable combination) and yet (ii) they can be made to be congruent by means of a mirror reflection.

The paradigmatic example of incongruent counterparts – which also is the one that Kant focuses on – is that of the left and right hands of our human bodies. A left glove cannot fit into a right hand. These material objects feature a difference in shape, as a result of which they have different handedness, which cannot be accounted for in purely relational terms, or so Kant believed. This is roughly because, given any of the arbitrary parts of the left hand and considering all the possible spatial relations holding between those parts, there are parts of the right hand that stand in exactly the same relation. This can be taken as an argument against relationalism. Kant goes one step further and writes:

> We desire to show that the complete ground of determination of the shape of a body [e.g. a hand] rests not merely upon the position of its parts relatively to one another, but further on a relation to universal space.[5]

This passage shows clearly that he was taking the argument of incongruent counterparts to provide not only an argument *against* relationalism, but also

---

[4]My reconstruction is indebited to Nerlich (1976), yet differs significantly from it in various respects.

[5]Kant (1768).

a direct argument *in favor of* substantivalism. This idea is usually neglected in the literature and will be made clear in my own reconstruction.

Before providing the aforementioned reconstruction let me briefly address one (minor) detail. The original argument starts off with a premise according to which every object that could have an incongruent counterpart would have a definite handedness. Imagine a possible world containing a single hand. That hand, according to Kant, would either be a left hand or a right hand. This last claim has been challenged in the literature.[6] I take this point for granted, but I think relationalists have more arrows to their quiver than simply denying this premise. Everything is in place in order to address a formulation of the argument.

(1) Every object that has a possible incongruent counterpart has a definite handedness;

(2) Every hand is either a left hand or a right hand;

(3) Any two objects (x, y) that are incongruent counterparts have different handedness;

(4) The left hand has a different handedness from the right hand (follows from 3);

(5) There is a spatial difference between the right and the left hand (namely their handedness) (follows from 4);

(6) According to the main tenet of relationalism, spatial difference should be accounted for in terms of different relations between material objects;

(7) According to the main tenet of relationalism, every spatial relation between material object boils down to:

   (i) spatial distances between those objects, and

   (ii) spatial angles between those objects;

(8) According to relationalism, the different handedness of the left and the right hand should be accounted for in terms of either:

   (iii) spatial distances between parts of the hands, and

   (iv) spatial angles between parts of the hands;

(9) For any two arbitrary parts (a, b) of the left hand such that D (a, b) is the distance from a to b, and for any three arbitrary parts of the left hand

---

[6]Especially in the relationalist-friendly literature. Indeed according to Kant the action of a creative cause in providing the right-handed would necessarily have to be different from the action of providing the left-handed. So Kant simply denies the possibility of relationalism because it will lead to *indeterminism*. On this point see Pooley (2003).

(a, b, c) such that A (a, b, c) is the angle between ab and bc, there are parts
a*, b*, c* of the right hand such that D (a*, b*) = D (a, b) and A (a*, b*,c*)
= A (a, b, c);

(10) There is no difference in spatial relation between parts of the hands
that could account for the difference in their handedness (follows from 9);

(11) C1: Relationalism cannot provide a satisfactory account of the spatial difference of incongruent counterparts (follows from 7, 8, 10).

This point, i.e. C1, is the first negative conclusion of Kant's argument.
It is a negative conclusion in that it is an argument against a particular
metaphysics, namely relationalism, and not a direct argument in favor of
another.

But as we saw in the passage above, Kant wishes to go further. Here is a
possible way to do so:

(12) The difference in handedness must be accounted for in more fundamental terms;

(13) There is "something" that grounds the difference in handedness (follows from 12);

(14) There must be another entity besides the hand (be it a right or left
hand), such that the difference in handedness is explained in terms of the
relations that those hands bear to that different entity (follows from 10, 13);

(15) (Universal) space provides the only ground to account for the difference in handedness, in that the left and right hand would differ in their
relation to space;

(16) C2: (*Universal*) space exists (follows from 14, 15).

Conclusion C2 is stronger than conclusion C1 in that it offers a positive argument in favor of a particular metaphysics, namely substantivalism, rather
than being just a negative argument against one. In the following section I
will provide different arguments used by the relationalist in order to counter
both C1 and C2.

3   *Tu Quoque* and Primitive Handedness: Some Possible Relationalist Responses to the Incongruent Counterparts Argument

The first argument I would like to put forward stems from some undeveloped
remarks in Earman (1989) and it is a classic *Tu Quoque* argument. Let us

concede for the sake of argument that there is something valid in conclusion C1. That conclusion makes a compelling point only insofar as the substantivalist has a better account for the difference in handedness[7] between the right and left hand. But has she? In what follows, I give some reasons to be skeptical about this last claim.

The major difference between the relationalist and the substantivalist is one of ontological commitments. Substantivalists can appeal to an independently existing space. I will make the simplifying assumption that this space is a collection of unextended spatial points and that every non-empty collection of points constitutes a spatial region. If such a space is admitted in one's ontology two different types of properties and relations are gained, namely properties or relations between spatial regions (be they point-like or not)[8] or relations between spatial regions and material objects located at those regions.

Analytic metaphysicians usually understand the latter notion along the following lines[9]: if a material object, x, is exactly located at a spatial region R, x shares the relevant geometrical properties with R, that is to say it has the same shape, occupies the same volume, has the same topological features and so on.

Now, consider the spatial regions at which the left and the right hand are respectively located. Apart from exotic counterexamples[10] it seems reasonable to assume that for any subregion of those regions there is a part of a hand that is exactly located there.

If so, it follows from how we understand exact location, that for any spatial regions $r_1, \ldots, r_n$ and for any (geometric) relation holding R such that $R(r_1, \ldots, r_n)$, there are parts $h_1, \ldots, h_n$ exactly located at $r_1, \ldots, r_n$ respectively, such that $R(h_1, \ldots, h_n)$ holds.

This lays the ground for the *Tu Quoque* argument, for it follows that for every relation between spatial regions there is a corresponding relationship between the material parts of a hand. Whatever account can be provided in terms of regions of space can (and should) be provided in terms of parts of material objects. Either the argument in (1)-(11) is somewhat flawed or it cuts both ways, in which case it undermines both relationalism and substantivalism. These seems however especially to undermine C1.

There seems to be a natural response to the *Tu Quoque* argument: it is that it unfairly restricts the substantivalist's considerations to the regions occupied by the hands. But the substantivalist's commitment to space goes well beyond that. It is a commitment to the entire space and it is the relationship of the hands to the entire space[11] that somehow ground their handedness.

---

[7]Which she might think she has, given C2. However, now we are simply focusing on the argument (1)-(11).

[8]The argument does not require her (i.e. the substantivalist) to take a stand on whether any spatial relations boils down to relations about point-like regions.

[9]For an authoritative introduction see Casati and Varzi (2009).

[10]For example the so-called extended simples, on which see Markosian (2014).

[11]Probably Kant's reference to a "universal space" can be taken as an indication that he actually already thought of this very possibility even if he was not able to develop it further. In modern terms the natural candidate for a general property of space that could

Isn't this the rationale behind the move from C1 to C2?

I believe that the relationalist is entitled to a further reply. We have just allowed the substantivalist a global property of space that goes well beyond the regions occupied by the hands. The relationalist should be allowed then at least a counterfactual appeal to an asymmetrical object that could be used to ground the different handedness of the hands.

In this new augmented relationalist framework the hands differ in their spatial properties insofar as, if an asymmetrical object were introduced in the universe, only one of them will enter in a particular relation (e.g. the thumb pointing in the same direction) with it.

Now, the substativalist can deny this counterfactual appeal. But I contend that this is not an appealing strategy. I cannot argue for this claim at length here but the point is roughly the following. Relationalism should be constructed as the claim that space is constructed out of possible spatial relations between more fundamental entities, namely material objects, on pain of being philosophically uninteresting.

An analogy with another example of the reductionist/anti-reductionist debate – the reductionists being the relationalists – might shed some light on what we mean here. Suppose I want to hold the reductive thesis that a whole is nothing over and above the sum of its parts. Naturally enough I have to concede to the reductionist that the relations holding between some parts should be taken into serious consideration, e.g. spatial (or causal) relations. If the reductionist is not allowed to consider relations between parts, her thesis becomes blatantly false and uninteresting: if I have four sticks and a wooden plank I do not yet have a table. The reductionist/anti-reductionist debate over composite objects is philosophically interesting only on the condition that some concessions are made to the reductionist. The same goes for the relationalist/substantivalist debate. The reductionist should be allowed to use modal language about possible locations. And this possibility is grounded in counterfactual claims.

But if so, relationalists could deny premise (15) and resist the argument for C2.

Before moving on to my conclusion, I would like to put forward another possible reply by the relationalist, one that undermines C1 and C2 at the same time.

Premise (7) of the argument for C1 constructs relationalism rather poorly. It maintains that relationalism is committed to reducing all spatial relations to distances and angles. But why should this be so? If substantivalists are allowed to augment their ontology, why should relationalists be denied to augment their own *ideology*? By way of an example the relationalist could add a new *primitive spatial determinable property*, namely handedness, which is able to differentiate between right and left handedness. Premise (7) should be rephrased accordingly in the following way:

(7*) According to the main tenet of relationalism, every spatial relationship

---

ground local spatial relations is orientability. But I will not pursue this line of argument here.

between material objects boils down to:

(i) spatial distances between those objects;

(ii) spatial angles between those objects, and

(iii) right or left handedness.

Naturally, the trouble for substantivalists is that the argument for C1 does not follow through with the premise of (7*), for consequently (10) turns out to be false. Nor does it follow through to the conclusion reached in C2, since that argument depends crucially on (10). Not only this, but (12) turns out to be false as well. If handedness is taken as a primitive feature, then it *should not be accounted for in more fundamental terms.*

4 Conclusion

In this paper I attempted to do two things, i) to offer a clear reconstruction of a neglected geometrical argument against relationalism; and ii) to offer some possible relationalist responses to this newly proposed reconstruction. It seems reasonable to say that geometrical considerations alone do not suffice to vindicate one metaphysics of space over another. But this is not to say that we should prefer one by virtue of some other considerations, which are not necessarily linked to the problem of motion. A quasi-geometrical argument in favor of substantivalism could probably be formulated by considering the global topological notion of orientability.

Such an argument would require further independent scrutiny. However, I would simply want to point out that it could be that orientability is not a purely geometrical property. For it might be the case that the property of orientability of space is rather defined by some law-like features that describe the behavior of material objects. The most promising strategy for this potential outcome is, naturally enough, to look for parity violations within fundamental particle physics.[12]

The nature of space is such a deep problem that we should have actually expected it not to be so easily solved. But aren't difficult problems also the most interesting?

References

Barbour, J. and Bertotti, B. (1982), "Mach'sprinciple and the structure of dynamical theories", London, Proceedings of the Royal Society, 1982.

Barbour, J. (1989), *Absolute or Relative Motion? A Study From the Machian Point of View of the Discovery and the Structure of Dynamical Theories.*

---

[12]For further information see Gardner (1960).

Cambridge University Press, 1989.

Kant, I. (1768), *Del Primo fondamento della distinzione delle regioni nello spazio*, in P. Cartabellese (trad. it.), *Scritti Minori*, Bari, Laterza, 1923, pp. 201-208.

Earman, J. (1971), "Kant, Incongruous Counterparts, and the Nature of Space and Space-Time", in *The Philosophy of Right and Left: Incongruent Counterparts and the Nature of Space* (The Western Ontario Series in Philosophy of Science), J. Van Cleve e R. E. Frederick (Ed.), Dordrecht, Boston, London, Springer Editori, 1991, pp. 131-150.

Earman, J. (1989), "On the other hand...: A reconsideration of Kant, Incongruent Counterparts, and the Absolute Space", in *The Philosophy of Right and Left: Incongruent Counterparts and the Nature of Space* (The Western Ontario Series in Philosophy of Science), J. Van Cleve e R. E. Frederick (Ed.), Dordrecht, Boston, London, Springer Editori, 1991, pp. 235-256.

Earman, J. (1989), *World Enough and Space-Time: Absolute versus Relational Theories of Space and Time*, Bradford Book, 1992.

Gardner, M. (1960), *The New Ambidextrous Universe*, New York, W. Penguin, 1990.

Markosian, N. (2014), "A Spatialapproach to Mereology", in Shieva Kleinschmidt (ed.), *Mereology and Location*, Oxford University Press, 2014, pp. 69-90.

Nerlich, G. (1973), "Hands, Knees, and Absolute Space", in *The Philosophy of Right and Left: Incongruent Counterparts and the Nature of Space* (The Western Ontario Series in Philosophy of Science), J. Van Cleve e R. E. Frederick (Ed.), Dordrecht, Boston, London, Springer Editori, 1991, pp. 151-172.

Pooley, O. "Handedness, Parity Violation, and the Reality of Space", in Brading and Castellani (eds) *Symmetries in Physics: Philosophical Reflections* (CUP, 2003).

4 ABOUT THE SUBSTANTIVAL NATURE OF SPACETIME

Anguel S. Stefanov

Abstract   The relationist-substantivalist debate about the nature of spacetime has not still come to an end. The aim of this paper is to present and analyze four arguments in support of the substantival nature of spacetime. The first one is an argument based on chirality. The second and the third one take into account the cosmological constant and the gravitational waves respectively, as displaying non-relational qualities of spacetime. And the fourth argument involves an arguable interpretation of the basic equation in the theory of general relativity, pretending to be the only consistent one. It is my claim that this fourth argument provides a general ontological framework for the validity of the first three arguments.

1   Introduction

Since the Newton-Leibniz debate about the substantival or relational nature of space and time the problem is still going to be a bone of contention among philosophers.

> There are two venerable traditions in the philosophy of space and time. One is 'substantivalism', which maintains that space and time (relativistically, spacetime) are objects that exist *in addition to* ordinary material objects such as tables and chairs. The opposing tradition, 'relationism', rejects the existence of space and time (spacetime) and maintains that all that exists is material objects. According to traditional relationism at each time there are spatial distances between material objects and there are temporal distances between events involving these material objects. [2, p. 125, my italics]

My paper will neither be based on historical facts about the debate, nor on the cogent arguments offered by Frank Arntzenius [2] in favour of sub-

A. S. Stefanov, M. Giovanelli (Eds), *General Relativity 1916 - 2016. Selected peer-reviewed papers presented at the Fourth International Conference on the Nature and Ontology of Spacetime, dedicated to the 100th anniversary of the publication of General Relativity, 30 May - 2 June 2016, Golden Sands, Varna, Bulgaria* (Minkowski Institute Press, Montreal 2017). ISBN 978-1-927763-46-9 (softcover), ISBN 978-1-927763-47-6 (ebook).

stantivalism. Its aim is the explication and the analysis of four relatively independent arguments supporting the substantival nature of spacetime. I say that these arguments are relatively independent not because they share some common theoretical parts, but because the fourth of them (the argument from a consistent interpretation of the basic equation in general relativity) may be considered to provide a general ontological framework for the first three.

2 Argument based on chirality

Incongruent counterparts are mirror objects, which, though being quite similar like the left and the right human hands still cannot be superimposed on each other through ordinary translations and rotations within a fixed orientable space. A left glove cannot fit a right hand, and vice versa. For the first time the possibility for the existence of such objects was taken as an argument for the existence of the absolute Newtonian space in the last precritical work of Kant from 1768 [6]. But if the argument is correct, and the absolute Newtonian space exists, then the three-dimensional physical space is substantival, and not relational.

If only one human hand existed in the world, Kant contended, it would be either a left, or a right hand. Relationists disagree, insisting that a single hand could neither be qualified as a left, nor as a right one, since there is no other object to be involved in relation to the solitary hand, so that its handedness to be ascertained. Let us imagine to this effect the existence of a suitable "referent object" alongside the lonely hand, and let the referent object be a handless human body. Then we may see to which of the two wrists of the body the hand will match. Suppose it matches the right wrist. Thus the right-handedness of the hand would be ascertained. But without the referent body no right-handedness of the solitary hand exists. Its righthandedness comes as a result of a relation to the handless body and is not a property of the hand itself, if it were the only existing thing in the universe.

There is a clear objection to this criticism. Let us revisit Kant's story with the solitary hand. It was accepted that it was the only existing thing in the universe, till the handless human body has come into being. The objection states that the hand was a right one even before the appearance of the referent object. Indeed, the appearance of the handless body does not affect the nature of the hand that was created before the body. The appearance of the latter does not affect the spatial characteristics of the region where the hand was situated, as well. But if so, then it certainly follows that the solitary hand was right for all of the time of its existence, and the referent body serves only for its right-handedness to be observed, and in no way to be created.

This objection supports the substantival nature of space, since it is an argument that space has something to do with the concrete handedness (chirality) of the hand. But what answer could be suggested to the question why a lone hand is either right or left *per se*. A general idea for such an answer was proposed for the first time in the same work of Kant [6, p. 20, original italics]:

> [M]y aim in this treatise is to investigate whether there is not to be found in the intuitive judgments of extension, such as are contained in geometry, an evident proof *that absolute space has a reality of its own, independent of the existence of matter, and indeed as the first ground of the possibility of the compositeness of matter.*

That absolute space, as a reality of its own, can be looked upon "*as the first ground of the possibility of the compositeness of matter*" is an original idea that has an explanatory potential. It implies the assumption that some inherent features of space (for instance the specific metric and topology of a space) could affect (at least) the in/congruency of geometrical objects. Thus incongruent geometrical figures drawn on a plane, to continue the story by flat images of left and right human hands, can become congruent counterparts, if placed on a Möbius strip, representing a non-orientable two-dimensional space. Kant was not in a position to develop further his original idea on the background of the classical physical and mathematical knowledge of his time. Nevertheless, he was not surprised that incongruent counterparts might display a functional asymmetry concerning some of their exhibited properties, like those of left and right human hands and ears [6, p. 30].

Having in mind the parity violation in the micro-world, Kant's argument could be extrapolated to the effect that if there were only one weak interaction breaking the CP-symmetry (the charge conjugation – parity symmetry) in the universe, it would do so. The difference now is that such a quantum process takes place in spacetime. But what is the role of time here? Its role is to restore the symmetry at a deeper level. A quantum physical system is invariant only with respect to the triple CPT transformation, including the operation of time reversal.

If spacetime had a relational nature, the last two statements could hardly be taken to be meaningful, because in a purely relational context they would have no reasonable explanation. Moreover, if spacetime had a relational nature, it would have no impact on the symmetry of physical interactions. Spatial distances and time intervals would be kept the same under rigid and symmetrical transformations, because spacetime is accepted to emerge *out of the relations* among material objects and force fields. But if it is true that symmetries are sometimes broken in isolation, or in couples, it comes out that spacetime affects the compositeness of matter (instead of being dependent on the latter), and thus has a substantival nature.

3 Argument concerning the cosmological constant

When A. Einstein firstly wrote his equation of the general theory of relativity, he introduced an additional term, known as the cosmological constant, so that the equation could describe a static Universe. When astronomic observations showed that this was not the case, he removed this term. However, 43 years after Einstein was gone, observations showed not only that the Universe is expanding, but that its expansion is accelerating. Contemporary cosmologists re-introduced Einstein's cosmological constant. They have done this for the sake of a consistent explanation for the observed acceleration of the expansion of the Universe. But even if this acceleration would not be confirmed by interpreting new astrophysical observational data, the universal expansion is an established fact, and it is certainly in need of an explanation.

> As well as matter, the universe may contain what is called "vacuum energy", energy that is present even in apparently empty space. . . vacuum energy causes the expansion to accelerate, as in inflation. In fact, vacuum energy acts just like the cosmological constant. . . that Einstein added to his original equations in 1917, when he realized that they didn't admit a solution representing a static universe. [5, pp. 96-97]

The energy ruling the expansion of the Universe, known by its popular name today as dark energy, is a fundamental quality that cosmologists refer to the "empty" spacetime itself. Dark energy opposes the effect of the universal gravitation that is empirically expressed by the well-known attractive force among material bodies. This force is inversely proportional to the square of the distances among material bodies, so that the gravitational interaction becomes weaker in an expanding space shifting material configurations aside from each other. However, if dark energy expressed by the cosmological constant is a quality of spacetime itself, its anti-gravitational effect ought to be one and the same independently of the fact how much the universal space has been expanded. Thus one may certainly expect that there must be a stage in the evolution of the Universe, when the effect of the dark energy would become stronger than the gravitational attraction. From this stage on the universal expansion would exhibit acceleration. And this is exactly what astronomers found to be the case in 1998.

> But then, as ordinary matter spread out and its gravitational pull diminished, the repulsive push of the cosmological constant (whose strength does not change as matter spreads out) would have gradually gained the upper hand, and *the era of decelerated spatial expansion would have given way to a new era of accelerated expansion.* [4, p. 300, his italics]

If the nature of spacetime were relational, then spacetime could hardly possess such an *intrinsic* dynamic quality as dark energy. Energy is a fundamental property of material systems, and they have an existence of their own. So, we must concede that spacetime, possessing energy of its own, has also an existence of its own; or in other words, it has a substantival nature.

4 Argument concerning gravitational waves

Since the birth of Einstein's general theory of relativity in 1916, it has been suggested that gravitational waves could exist. They are ripples in the curvature of spacetime that propagate as waves at the speed of light. One hundred years after Einstein hypothesized their existence, on February 11, 2016, the LIGO Scientific Collaboration and Virgo Collaboration teams (covering the international participation of scientists from several universities and research institutions) announced that they had made the first observation of gravitational waves. They originated from a pair of merging black holes being at a distance of 1.3 billion light years from the Earth, somewhere beyond the Large Magellanic Cloud in the southern hemisphere sky.

> The discovery is a great triumph for three physicists — Kip Thorne of the California Institute of Technology, Rainer Weiss of the Massachusetts Institute of Technology and Ronald Drever, formerly of Caltech and now retired in Scotland — who bet their careers on the dream of measuring the most ineffable of Einstein's notions. [7]

I will not comment here how precisely the experiment was carried out, although the history of its planning and realization deserves a special attention. As far as I am aware, its positive result has been accepted by the scientific community. Besides, the LIGO – VIRGO Scientific Collaboration teams announced on June 15, 2016, that a second detection of gravitational waves from coalescing black holes was observed.

What is important here for my purpose is the following. If the gravitational waves could not be detected for some principal reason, then this would be no good news for the proponents of the substantival view of spacetime. But what after their existence was confirmed?

The observation of gravitational waves represents a clear argument in support of substantivalism. Relationism could hardly account for the existence of such waves. Indeed, from a relationalist point of view, only material objects really exist, while space and time are specific relations among them. However, relations are relational *properties* of objects, and as such properties they have no existence of their own. But if so, relational properties cannot possess non-relational properties on their part, and in particular, spacetime cannot initiate gravitational waves, as being a genuine disturbance of spacetime, even if they transmit no energy. On the contrary, only if spacetime exists as an entity of its own and exhibits local curvatures responsible for the gravitational interaction, then collisions of massive cosmic objects like galaxies and black holes can certainly account for the appearance of gravitational waves.

5   Argument from a consistent interpretation of the basic equation in general relativity

I have in mind the so called by A. Einstein field equation, or his well-known tensor equation of the general theory of relativity (the one hundred anniversary of the publication of which we are celebrating this year):

$$R_{\alpha\beta} - \tfrac{1}{2} g_{\alpha\beta} R = \kappa T_{\alpha\beta}.$$

As is well known, the left side of this equation is usually called now Einstein's tensor, and it refers to the geometry of spacetime, but more ontologically speaking, to the entire set of spatial-temporal events. The tensor at the right side is the tensor of matter, known also as the energy-momentum tensor, and is taken to structurally represent the state and distribution of the different kinds of matter. However, Einstein himself had a problem concerning the construal of his field equation [3, p. 370]:

> But, it is similar to a building, one wing of which is made of fine marble (left part of the equation), but the other wing of which is built of low grade wood (right side of equation). The phenomenological representation of matter is, in fact, only a crude substitute for a representation which would correspond to all known properties of matter.

At that, there is another interpretative problem concerning the motion of matter according to the general theory of relativity:

> The theory incorporates the effect of gravity by saying that the distribution of matter and energy in the universe warps and distorts spacetime, so that it is not flat. Objects in this spacetime try to move in straight lines, but because spacetime is curved, their paths appear bent. They move as if affected by a gravitational field. [5, p. 35]

So, we are faced with a *curious situation*: material bodies warp spacetime, while at the same time spacetime curvatures determine the movement of material bodies.

The just outlined problems point to the need of a consistent interpretation of Einstein's basic equation of general relativity.

As it seems, there are two interpretative possibilities. The first one is to construe the equation as a standard equality of two different kinds of tensors, representing *independent kinds of entities* – Einstein's tensor and the matter tensor (referring to spacetime and the composition of matter, respectively). At that, the tensor of matter is of a primary significance, since it is said that material bodies do cause the curvature of spacetime. In this case, however, the curious situation at hand could not be consistently elucidated. That is to say, this interpretation provides no arguable answer to the questions "Why, and how material objects warp spacetime?"

The remaining alternative is to construe the equation as *expressing an identity*, and not merely a correlation of equality between its left and right sides. Thus both these sides ought to be taken as theoretical constructs that refer to one and the same entity. It is certainly represented by the "fine marble (left part of the equation)", or in other words, this initial entity is spacetime.

Still there is one more reason in favour of the identity interpretation, and it is of a logical character. As is well known, the covariant derivative of the tensor at the right part of the equation – the tensor of matter, or the energy-momentum tensor – must be zero. Applying covariant derivation includes the Christoffel symbols of the second kind (which are the affine connections of the four dimensional Riemannian spacetime). The Christoffel symbols, however, are functions of the metric tensor and its ordinary derivatives. Thus it comes out that in order to see whether the tensor at the right side of the equation is really a tensor of matter, one has to know beforehand the metric of the spacetime. This vicious circle could be overcome only by the identity interpretation, since within it spacetime and matter (or better say spacetime without and with material bodies) belong to one and the same initial, or fundamental essence.

What is this fundamental essence?

It has been shown that according to a consistent reconstruction of the general theory of relativity the concept of spacetime as a world of physical events has a logical priority to the tensor of matter [1, p. 250]. At the same time the identity interpretation takes the referents of the tensor structures at both sides of Einstein's basic equation to be, or fall into one and the same, ontological essence. The latter then must be identified somehow with space-time, but not only with the geometry of spacetime. It would not be correct the geometry of spacetime and matter to be separated as two independent entities. On the contrary, they must be construed as two cognitively separable parts of a unique ontological essence. This could certainly be neither "empty" spacetime, i.e. spacetime without matter, nor "pure" matter without spacetime (that could even hardly be conceived of). It could be provisionally named "prime-matter", or "primal matter", and so to remind us of the ancient Greek idea of a prim(aev)al essence giving birth to the variety of all visible and tangible natural objects, or of something like Anaximander's apeiron. When Einstein's tensor equals zero, then prime-matter is reduced to "empty" Riemannian spacetime; and when it is different from zero, then prime-matter presents itself as spacetime filled with material structures. Prime-matter is the fundamental essence that is looked for, since it unites spacetime as an entity described by a mathematical language with the material structures emerging within it.

The identity interpretation provides an ontological framework for the three previous arguments for the substantival character of spacetime. Space-time, interpreted as prime-matter, is the genetic background for the emergence and the compositeness of matter. And it also possesses an immutable feature of matter – energy of its own. This is not strange at all, since according to the suggested interpretation spacetime – accepted as prime-matter –

is the fundamental element in the theory, while all the properties and interactions of material objects are taken to be specific states of prime-matter.
Gravitational waves as ripples in the fabric of spacetime are properties of this
same prime-matter.

The final conclusion in the end is that the identity interpretation of the
Einstein's equation of general relativity certainly excludes the possibility for
the relational nature of spacetime. It could be thought in no way as some set
of relations among material objects whatsoever, because, just on the contrary,
it is spacetime in its quality of prime-matter, which gives birth to material
structures, and not vice-versa.

Space-time has a substantival nature even not in the traditional sense of
this qualification. It was stated at the beginning in Arntzenius' words that
traditional substantivalism "maintains that space and time (relativistically,
spacetime) are objects that exist *in addition to* ordinary material objects
such as tables and chairs." As we have seen, however, it could be said that
spacetime does not merely exist in addition to material objects; it is the very
base for their existence.

References

[1] Anastassov, Anastas. 1973. "On the Logical Structure of Physical
Theories and Particularly of the Relativity Theory (Space, Time, Matter)"
(in Bulgarian) In (A. Polikarov et al eds.), *Contemporary Physics. Directions
of Development, Methodological Problems.* Sofia: "Nauka i Izkustvo", 241-262.

[2] Arntzenius, Frank. 2012. *Space, Time and Stuff.* Oxford, New York:
Oxford University Press.

[3] Einstein, Albert. 1936. "Physics and Reality." *J. F. I.*, March,
349-382. (Translation by Jean Piccard.) Hosted by Prof. M. Kostic at:
www.costic.niu.edu

[4] Greene, Brian. 2004. *The Fabric of the Cosmos. Space, Time, and
the Texture of Reality.* New York: Alfred A. Knopf.

[5] Hawking, Stephen. 2001. *The Universe in a Nutshell.* London, New
York, Toronto, Sydney, Auckland: Bantam Press.

[6] Kant, Immanuel. 1991. "On the First Ground of the Distinction of
Regions in Space." In *The Philosophy of Right and Left. Incongruent Counterparts and the Nature of Space*, edited by James Van Cleve and Robert E.
Frederick, 27-33. Dordrecht / Boston / London: Kluwer Academic Publishers.

[7] Overbye, Dennis. 2016. "Gravitational Waves Detected, Confirm-
ing Einstein's Theory", http://www.nytimes.com/2016/02/12/science/ligo-
gravitational-waves-black-holes-einstein.html?_r=0

5 THE MEANING OF SPACETIME SINGULARITIES

OVIDIU CRISTINEL STOICA

Abstract A recent extension (not modification) of semi-Riemannian geometry and general relativity has been proven to work for a large class of singularities, which includes the major known ones, and to provide a description of them in terms of finite invariant geometric and physical objects. An interesting consequence is that one can no longer conclude that the existence of these singularities represents the breakdown of general relativity. After a brief review of these results, some implications on the nature and ontology of spacetime are discussed, in particular at singularities. The proposed approach suggests that it is relevant to understand what geometric and physical objects are more fundamental and why. In order to achieve this goal, the underlying mathematical structures of spacetime are deconstructed. In particular, the very notion of metric, connection, curvature, causal structure, stress-energy tensor, are revised. The analysis suggests that the structure of lightcones plays a more fundamental role than other structures.

1 Introduction

In this article I will discuss the geometric and physical interpretation of spacetime singularities occurring in *general relativity*.

General relativity has two main problems: the prediction of *singularities* [22], and the problem of *quantization*. Despite these problems, the predictions of general relativity continue to be confirmed by experiment, culminating recently with the detection of gravitational waves resulting from the merging of two black holes [18].

Given the repeated experimental confirmation of the predictions of general relativity as compared to the alternative theories, we should consider more carefully what general relativity itself has to say about singularities. This motivated my research program to find a natural formulation of general relativity in terms of variables that remain finite at singularities, *cf.* [33] and references therein. I will briefly review these results, and then discuss the ge-

A. S. Stefanov, M. Giovanelli (Eds), *General Relativity 1916 - 2016. Selected peer-reviewed papers presented at the Fourth International Conference on the Nature and Ontology of Spacetime, dedicated to the 100th anniversary of the publication of General Relativity, 30 May - 2 June 2016, Golden Sands, Varna, Bulgaria* (Minkowski Institute Press, Montreal 2017). ISBN 978-1-927763-46-9 (softcover), ISBN 978-1-927763-47-6 (ebook).

ometric and physical interpretation of singularities and singular spacetimes, as well as the implications on the nature and ontology of spacetime.

2 Spacetime singularities

In general relativity the metric tensor is dynamic, and so are the Levi-Civita connection (needed for covariant derivatives), the geodesics, and the Riemann curvature. Einstein's equation specifies how the geometry and the matter fields evolve in an interdependent way. The metric is a symmetric tensor, specified at each point, in a coordinate system, by a symmetric 4×4 matrix. The entries g_{ab} of the matrix are dynamical quantities. Both geometry and the field equations are based on the metric, and they can work in their usual form only if we assume that the determinant of g_{ab} never vanishes, and that each component g_{ab} is finite, and that their first and second order partial derivatives are finite too. However, there is no guarantee that this always happens. The dynamics changes the components g_{ab} of the metric sometimes violently, and there is no way to be sure that the determinant will never vanish, and that none of its components will never become infinite. When any of these two possibilities happens, the metric is *singular*.

The first exact solutions of Einstein's equation showed that this may happen. Both the Schwarzschild black hole solution [26, 25] and the big-bang cosmological model of Friedmann-Lemaître-Robertson-Walker (FLRW) [10, 11, 17] have singularities. While this was hoped to be a special case due to the too high degree of symmetry of the solutions, it was proven, through the *singularity theorems* by Penrose [22, 23] and Hawking [12, 13, 14, 15], that it is actually much more general, and unavoidable under reasonable conditions.

The problem with singularities is that the metric becomes singular. This means that some of the metric tensor components g_{ab} or g^{ab} become infinite. This prevents the construction of the covariant derivative $\Gamma^a{}_{bc}$ (since $\Gamma^a{}_{bc}$ requires the inverse of the metric) and the Riemann curvature $R^a{}_{bcd}$. However, in [37] I show that differential geometry can be extended in a natural and invariant way to singular metrics g_{ab} which are smooth and become degenerate ($\det g = 0$).

3 Does spacetime break down at singularities?

Is the prediction of singularities the omen of the breakdown of general relativity? I will explain that it is not the case, and that only the usual equations we use to understand them have this problem. But it is possible to change the equations, not by modifying them, but in a way similar to a change of variables. The resulting equations are expressed in terms of variables that do not blow up at singularities, and outside the singularities the solutions coincide with the standard ones. The new variables are as natural as the usual ones and from certain points of view more fundamental, both geometrically and physically.

In [37, 43, 34] I studied metric singularities for which the components g_{ab} remain finite, and the determinant of the metric vanishes. Let us call such singularities *benign*, and let us call *malign singularities* those for which g_{ab} blows up for some components. The result was a generalization of semi-Riemannian geometry, which I applied in subsequent articles to the spacetime singularities in general relativity (see [33] and references therein). For a large class of benign singularities, one can still have geometric and physical objects that remain finite at the singularities, which satisfy invariant field equations equivalent to the usual ones outside the singularities.

Even for the benign singularities, those for which the metric is smooth but the determinant vanishes, one cannot construct the covariant derivative and the Riemann curvature as usually. The reason is that $\Gamma^a{}_{bc}$ and $R^a{}_{bcd}$ are constructed using the inverse of the metric, g^{ab}, which blows up when $\det g_{ab} = 0$.

But the *lower covariant derivative* (which can be expressed in terms of the *Christoffel symbols* of the first kind Γ_{abc}) and the lower-index form of the Riemann curvature R_{abcd} remain finite at such singularities. This turned out to be enough to describe a large class of singularities, and to rewrite Einstein's equation in terms of quantities that remain finite, and the solutions are still equivalent to the solutions of the original Einstein equation outside the singularities [37, 35]. This applies to FLRW and more general big bang solutions [41, 30, 32].

Although for the malign singularities the problem is a bit more difficult, there is a solution for their case as well. All examples of stationary black holes contain malign singularities, but they can be reduced to benign singularities by using singular coordinate transformations, similarly to the case of the event horizon, which was resolved by Eddington [8] and Finkelstein [9]. Of course, unlike the case of the event horizon, for which there are singular coordinate transformations that remove the singularity, for the $r = 0$ singularities this will not work to remove it, because the *Kretschmann invariant* $R_{abcd}R^{abcd}$ blows up. But the singularity can be made benign by such transformations, and then the mathematical apparatus I developed in [37, 43, 34] can be applied. In the following I will detail this approach.

The fact that black hole singularities are apparently not benign, but malign, may be explained by considering that the usual coordinates are themselves singular, similarly to the case of the event horizon, which was resolved by Eddington and Finkelstein. The singular coordinate transformation I used, similar to their method, could be used to make the $r = 0$ singularity of black holes smooth, albeit degenerate [31, 29, 40]. The mentioned methods developed for degenerate metrics with benign singularities can then be applied, and the Schwarzschild solution can be extended analytically beyond the singularity. Singularities turn out to be compatible with global hyperbolicity [31, 38], allowing thus the conservation of information during black hole evaporation.

Not only that the singularities in general relativity turned out to be understandable in terms of finite quantities, but they may also provide a solution to the problem of quantum gravity. The geometric understanding of singularities from this approach leads to the conclusion that they are accompanied by *dimensional reduction* effects, which are researched in the last years because they allow the removal of infinities in *perturbative quantum gravity*. Many, perhaps most of the approaches to quantum gravity have something in common – they either imply, or rely on dimensional reduction [7, 6]. Usually the assumptions of dimensional reduction, either direct or indirect, appear to be made *ad-hoc*, in order to allow the perturbative renormalizability of quantum gravity. However, several of these approaches are supported by the very geometric properties of the singularities. In perturbative expansions in terms of point-like particles, particles become tiny black holes, and the singularities weight out the amplitudes in such a way that dimensional reduction effects are possible [36].

I will now discuss the implications on what mathematical and physical objects are more fundamental for spacetime.

4   The mathematical structure of spacetime

In general relativity, spacetime is considered to be a differentiable manifold, endowed with a Lorentzian metric. This assumes implicitly an entire hierarchy of mathematical structures, and it is not easy to answer the question which are the most fundamental.

Think for example at the Euclidean plane. One may consider that the notions of distance between two points and of angle between two lines are fundamental. On the other hand, we can have already a well defined mathematical structure even without the notions of distance and angle, based only on the axioms specifying the relations between points and lines. The *affine geometry* of the Euclidean space thus relies only on the notions of points and lines, without appealing to a metric. This allows one for example to see that the affine structure of a four-dimensional Euclidean space is identical to that of the Minkowski spacetime. The difference between them is introduced by the metric. The metric notions such as distance and angle are introduced by the *congruence axioms* [16]. However, the fact that there is a rich structure already which does not rely on the metric does not prove that the Euclidean metric is less fundamental than the lines. We can proceed in a different order, and start with a *metric space*, which is a set of points endowed with a distance between pairs of points. From this, we can define a topology, and the geodesics as those continuous curves of minimal (or extremal, in general) length. Sometimes it is easier to consider the affine structure as more fundamental, while other times the metric.

More generally, from mathematical point of view, a differentiable manifold endowed with a metric, in particular the spacetime, is also a hierarchy of mathematical structures. First, the spacetime consists of events, which form a set. The set is endowed with a topology. Then, each open set of the spacetime is required to be *homeomorphic* (topologically equivalent) with an open set of an Euclidean space, this endowing the spacetime with a dimension and a structure of *topological manifold*. Since the fields satisfy equations with partial derivatives, one needs to add a *differential structure* on top of this.

Usually, the metric tensor, which gives the geometry, is introduced as the next structural level. Then, with the help of the metric, one defines a *Levi-Civita connection*, which has the role of describing the parallel transport, and from which one derives the geodesics and the Riemann curvature tensor. Then, the Einstein equation is an equation relating the curvature with the distribution of matter fields.

But geometers know that a connection does not necessarily require a metric, and it can still be used to define geodesics and a curvature. Moreover, geodesics can simply be a collection of lines, with no reference to differentiability, connection, or even length.

The metric of a general relativistic spacetime is defined at each spacetime event by 10 parameters. The metric at each event can be recovered only from the structure of lightcones, or from the null geodesics, up to a scaling factor [20]. This holds for *distinguishing spacetimes*, which cover the physically reasonable spacetimes (for instance the condition to be distinguishing rules out the closed timelike curves). Thus, the *horismos relation* (two events are in the horismos relation if and only if there is a null geodesic joining them) seems to be more fundamental than the metric, although we normally define this relations using the metric. But the fact that we can just start from a set endowed with a generic reflexive relation, which is considered to be the horismos relation, and recover most properties of a spacetime, shows that it is indeed possible that this relation is more fundamental than the metric [42]. This works for any reflexive relation, and the spacetime can even be discrete. Thus, the structure of lightcones, or the causal structure of spacetime, may be more fundamental than the metric. This plays an important role in the interpretation of the singularities that follows from the approach discussed here.

5   What mathematical formulation is more fundamental for spacetime?

If Nature prefers to use the proposed variables and atlases, it has to do this not just as a trick to avoid the infinities at singularities, but for more fundamental geometric and physical reasons. In the following, I try to elucidate these reasons.

Spacetime has a *topological*, a *differential*, and a *(geo)metric structure*, built one in top of another. The more fundamental are considered to be the topological and the differential structures. The metric is a dynamical quantity, which depends on the stress-energy of matter. Being dynamical, there is nothing to stop it from becoming degenerate at some places, and this is why singularities appear. The fact that the metric is less fundamental than the manifold structure agrees with our mathematical understanding of differential geometry. However, physically, it is possible that the *causal structure* (representing the type of intervals separating the spacetime events) is more fundamental than the differential structure. This view is supported by the fact that the topology of lightcones is not affected at the important big bang and black hole singularities, while their differential structure is affected [39]. Thus, the structure of lightcones, or the causal structure of spacetime, may be more fundamental than both the metric and the differential structure. In [39] I showed that the topology of lightcones remains intact at the known black hole and big bang singularities, and only their differential and metric structures differ from those of a Minkowski lightcone.

Another question is related to the connection and the curvature. The connection specifies isometries between the tangent spaces at infinitesimally closed events. If the lower connection is more fundamental, it should also admit a geometric interpretation. The lower connection, rather than connecting the tangent spaces, connects the tangent space at an event with the cotangent space at an infinitesimally closed event in spacetime. Its non-commutativity is expressed by the lower Riemann curvature R_{abcd}, which may be more fundamental, if we think that this tensor and not $R^a{}_{bcd}$ exhibits the known symmetries at permutations of indices, the decomposition in the Weyl and Ricci curvatures, and the corresponding spinorial decomposition.

6   What physical objects are more fundamental for spacetime?

Regarding the physical content, the proposed replacement of Einstein's equation is

$$R_{ab}\mathrm{d}_{vol} - \frac{1}{2}g_{ab}R\mathrm{d}_{vol} + g_{ab}\Lambda\mathrm{d}_{vol} = \frac{8\pi G}{c^4}T_{ab}\mathrm{d}_{vol},\qquad(1)$$

which is clearly equivalent to Einstein's outside the singularities, where $\mathrm{d}_{vol} \neq 0$, but its terms remain finite at singularities at least in some important cases. Is $T_{ab}\mathrm{d}_{vol}$ more fundamental than T_{ab}? It should be, considering that what we integrate in order to obtain the mass or the momentum are the volume forms of the form $T_{ab}u^a u^b \mathrm{d}_{vol}$. This is clear for example if the stress-energy corresponds to a fluid, $T_{ab} = (\rho + p)u_a u_b + pg_{ab}$. One integrates the differential forms $\rho\mathrm{d}_{vol}$ and $p\mathrm{d}_{vol}$, and not the quantities ρ and p, which are not invariant, depending on the coordinates. While physicists think of them as scalar quantities, they are as scalar as the coordinates, and are defined in terms of a particular coordinate system. The truly invariant quantities are the differential forms $\rho\mathrm{d}_{vol}$ and $p\mathrm{d}_{vol}$. This is consistent with the fact that on differentiable manifolds mathematicians integrate volume forms, not scalar or tensor fields. In addition, the Lagrangian density is $R\mathrm{d}_{vol}$, and the corresponding equation is (1) rather than the usual Einstein equation, which is obtained by dividing (1) by d_{vol}, operation prohibited when the metric is degenerate, because $\mathrm{d}_{vol} = 0$, and it leads to infinities. In the particular case of the FLRW spacetime, the quantities $\rho\mathrm{d}_{vol}$ and $p\mathrm{d}_{vol}$ remain finite in the Friedman equations [41].

The above considerations suggest that the quantities used in rephrasing the geometry and physics to work at singularities are at least as adequate as the standard ones, both from physical and from geometric points of view.

7 Could spacetime be emergent?

The discussion we had so far about which mathematical and physical structures are more fundamental for spacetime assumes that spacetime is a continuum, as we understand it from general relativity. This does not exclude, however, the possibility that spacetime is discrete. There are some indications that point towards the idea that at least the information in a spacetime region has an upper bound, given by the Bekenstein-Hawking bound [5, 4]. Another argument goes along the line that since quantum mechanical systems are quantized, spacetime must be quantized as well. While this is true, the stronger argument that quantized means discrete, and spacetime therefore has to be discrete, is too far-fetched. It probably has to be true that spacetime itself is quantized in one form or another, but the argument itself is not rigorous, because even the quantum states of the Hydrogen atom are discrete only in what concerns the energy – the wavefunction of the electron in the bounded states spreads continuously throughout the entire spacetime.

In other words, when one says that a quantum system is discrete, one has to specify in which domain we consider the spectrum. When we say that the Hydrogen atom has discrete energy levels, we talk about the *frequency domain*, hence of the energy spectrum. This has no implications of discreteness on the position domain. By contrary, in the *position domain* there is no shred of discreteness, the wavefunctions are not even localized in a finite region of space. This is made clear by the uncertainty principle.

As an analogy, consider the two main types of digital graphic formats. One way is to store images as pixels, and the other way, as vector graphics. Both are constrained physically to store only a finite amount of information, but one of them is discrete in space, and the other one it is not. The vector graphics formats have infinite resolution, and the image at each possible scale is calculated using geometric formulae of lines, splines, and other geometric figures. So digital does not necessarily mean discrete in space. Similarly, spacetime can be such that the information enclosed in a finite region may be finite, without the spacetime itself being discrete.

But nevertheless there are promising approaches to quantum gravity in which spacetime itself is discrete, such as *Causal Sets* [27, 28], *Causal Dynamical Triangulations* [1, 2], *Loop Quantum Gravity* (LQG) [24], *emergent gravity* [45, 44] *etc.* The possibility that spacetime itself is discrete is currently under active consideration, perhaps more than ever.

If spacetime itself is discrete, it may seem easier to impose conditions that avoid the singularities. For example, the Einstein equations are not really a limit of those of LQG, but merely an approximation. For example, in *Loop Quantum Cosmology* it is easy to impose conditions that remove the big bang singularity [3]. These conditions are not compatible with the hypothesis of the singularity theorems in general relativity, and this is why it is possible to avoid the occurrence of singularities.

However, even if it is the case that spacetime itself is discrete, and it might very well be, the results presented here about the structure of singularities can turn out to be useful. It may still be possible to have degenerate metrics, and hence singularities of this kind. Or, if we remove them by some condition in some discrete spacetime theory of quantum gravity, at least the singularities will correspond in that theory to situations where the arrangement of the spacetime "atoms" is special, in the sense that it is highly degenerate or extreme in some directions. For instance, the big bounce from Loop Quantum Cosmology has a "bottleneck" of highest curvature and smallest radius, where the bounce happens.

The experimental tests of general relativity show that the theory, even if it would not be completely accurate, is at least a surprisingly good approximation of reality. This is usually explained by assuming that the discreteness is manifest at very small scales. But there are other reasons that ensure this. In [42] I showed that, no matter whether the spacetime is discrete or not, or maybe a hybrid between the two, one can derive properties of spacetime, like a topology, geodesics, dimension, and the metric up to a scaling factor, from the horismos relations alone, with minimal additional constraints mainly in the case of dimension. So the conceptual content of general relativity is something that remains stable and goes far beyond the continuous-discrete dichotomy. And the fact that the horismos relation, or the causal structure, are so fundamental, is also a lesson we learn from the approach I proposed to spacetime singularities [39].

8 Conclusions

I explained that it is possible to formulate the equations of general relativity in a way which is equivalent to the standard way outside the singularities, but in addition can be applied at the singularities, where they yield finite quantities. The geometric and physical objects that I used in this approach are invariant and natural. The fact that the proposed formulation appears to work in regimes in which the usual formulation does not work raises questions like: "Are there other reasons to accept the proposed formulation? The standard formulation of general relativity appeared to be natural both from differential-geometric and physical reasons. What if the alternative formulation proposed here is unnatural?".

Such questions are legitimate, and I addressed them in this article. We have seen that even for the Euclidean geometry of the plane, the mathematician does not always have a reason to consider one of the structures more fundamental than another. The choice is usually dictated by the interest in a particular structure, and by the applications. Mathematicians can take any of the structures in the hierarchy and abstract them. The notion of *forgetful functor* from *category theory* [19] allows ones to move from the category of mathematical structures of a kind to the category of mathematical structures of other kind, while ignoring or forgetting some of the mathematical structures. This abstraction does not have a prescribed order of forgetting, but rather there are more paths, and the different orderings of abstractions commute. So how do we know which of the mathematical structures is more fundamental in general relativity?

It appears natural to consider the most general structures as being more fundamental. With this choice, the lower covariant derivative and the lower Riemannian curvature appear to be more fundamental, since they apply to more situations. The usual covariant derivative and Riemann curvature obtained from a non-degenerate metric make sense only for such metrics, while the ones from the proposed approach make sense to more general metrics, which include the degenerate case and the main known singularities in general relativity.

From physical point of view, as seen from the example of the FLRW spacetime, the densitized formulation is more natural, since the energy and pressure densities ρd_{vol} and $p d_{vol}$ correspond to actual densities, while their scalar counterparts ρ and p are dependent on the coordinates in which $\det g_{ab} = 1$, and are not densities, being scalars. When performing integration, the densities are the actual integrand, while in order to integrate scalars, one needs to add the volume form as a correction for the integral to make sense, as it is known from any textbook of analysis on manifolds [21].

However, while these arguments allow us to select the more adequate geometric and physical quantities, they have no implication on the choice of the prefered atlas or differential structure. In this case, in the spirit of what Eddington and Finkelstein did, I suggest it is preferable to choose the atlas which allows the quantities to remain finite. I think this argument leaves room for improvement even in their case, because we do not have a mathematical result which shows that this choice is unique. It may very well be possible that a different choice leads to finite solutions, yet the solutions, when extended beyond the singularity, may be different. In the case of the Schwarzschild solution [31], an infinity of possible singular coordinate transformations are available to make the Schwarzschild metric degenerate. However, only one of the transformations I found results in a semi-regular metric in the sense of [37]. This hints towards the possible existence of a criterion for selecting the fundamental atlas, which still remains to be found.

A relevant hint in the direction of finding this criterion follows from the importance of the causal structure of spacetime, in particular of the horismos relation. The lightcone structure is strongly distorted at singularities, but since the lightcones preserve at least their topological structure, this indicates that they may be more fundamental not only than the metric, but even than the differential structure, as explained in [42, 39]. So the condition that the topology of the lightcones remains intact at singularities may be a relevant condition which allows us to select the correct atlas in which the metric tensor to be made benign or, if possible, non-singular, even if in the original atlas it appeared to be malign.

These discussions suggest that it is time to reconsider the ontology of spacetime in general relativity, and we accumulated several new tools that allow us to do this in a deeper conceptual way. Spacetime has still much to teach us, and its nature and ontology should always remain an object of critical investigation, as it is relevant not only for the philosophical foundations, but also as guideline for the new theories of quantum gravity.

References

[1] J. Ambjørn, J. Jurkiewicz, and R. Loll. Emergence of a 4D world from causal quantum gravity. *Phys. Rev. Lett.*, 93(13):131301, 2004.

[2] J. Ambjørn, J. Jurkiewicz, and R. Loll. Quantum gravity, or the art of building spacetime. In Daniele Oriti, editor, *Approaches to Quantum Gravity: Toward a New Understanding of Space, Time and Matter*, pages 341–359. Cambridge University Press, 2009.

[3] A. Ashtekar. Singularity Resolution in Loop Quantum Cosmology: A Brief Overview. *J. Phys. Conf. Ser.*, 189:012003, 2009.

[4] J. M. Bardeen, B. Carter, and S. W. Hawking. The four laws of black hole mechanics. *Comm. Math. Phys.*, 31(2):161–170, 1973.

[5] J.D. Bekenstein. Black holes and entropy. *Phys. Rev. D*, 7(8):2333, 1973.

[6] S. Carlip. The Small Scale Structure of Spacetime. *http://arxiv.org/abs/1009.1136*, 2010.

[7] S. Carlip, J. Kowalski-Glikman, R. Durka, and M. Szczachor. Spontaneous dimensional reduction in short-distance quantum gravity? In *AIP Conference Proceedings*, volume 31, page 72, 2009.

[8] A. S. Eddington. A Comparison of Whitehead's and Einstein's Formulae. *Nature*, 113:192, 1924.

[9] D. Finkelstein. Past-future asymmetry of the gravitational field of a point particle. *Phys. Rev.*, 110(4):965, 1958.

[10] A. Friedman. Über die Krümmung des Raumes. *Zeitschrift für Physik A Hadrons and Nuclei*, 10(1):377–386, 1922.

[11] A. Friedman. Über die Möglichkeit einer Welt mit konstanter negativer Krümmung des Raumes. *Zeitschrift für Physik A Hadrons and Nuclei*, 21(1):326–332, 1924.

[12] S. W. Hawking. The occurrence of singularities in cosmology. *P. Roy. Soc. A-Math. Phy.*, 294(1439):511–521, 1966.

[13] S. W. Hawking. The occurrence of singularities in cosmology. II. *P. Roy. Soc. A-Math. Phy.*, 295(1443):490–493, 1966.

[14] S. W. Hawking. The occurrence of singularities in cosmology. III. Causality and singularities. *P. Roy. Soc. A-Math. Phy.*, 300(1461):187–201, 1967.

[15] S. W. Hawking and R. W. Penrose. The Singularities of Gravitational Collapse and Cosmology. *Proc. Roy. Soc. London Ser. A*, 314(1519):529–548, 1970.

[16] D Hilbert. *The foundations of geometry*. Open Court Publishing Company, La Salle, Illinois, 1950.

[17] G. Lemaître. Un univers homogène de masse constante et de rayon croissant rendant compte de la vitesse radiale des nébuleuses extragalactiques. *Annales de la Societe Scietifique de Bruxelles*, 47:49–59, 1927.

[18] LIGO and VIRGO collaborations. Observation of gravitational waves from a binary black hole merger. *Phys. Rev. Lett.*, 116:061102, Feb 2016.

[19] S. Mac Lane. *Categories for the working mathematician*, volume 5. Springer Verlag, 1998.

[20] E. Minguzzi. In a distinguishing spacetime the horismos relation generates the causal relation. *Classical and Quantum Gravity*, 26(16):165005, 2009.

[21] R. Narasimhan. *Analysis on Real and Complex Manifolds*. Masson & Cie, Paris, 1973.

[22] R. Penrose. Gravitational Collapse and Space-Time Singularities. *Phys. Rev. Lett.*, 14(3):57–59, 1965.

[23] R. Penrose. Gravitational Collapse: the Role of General Relativity. *Revista del Nuovo Cimento; Numero speciale 1*, pages 252–276, 1969.

[24] Carlo Rovelli. Loop quantum gravity. *Living Rev. Rel*, 1(1):41–135, 1998.

[25] K. Schwarzschild. Über das Gravitationsfeld eines Kugel aus inkompressibler Flüssigkeit nach der Einsteinschen Theorie. *Sitzungsber. Preuss. Akad. d. Wiss.*, pages 424–434, 1916. *http://arxiv.org/abs/physics/9912033*.

[26] K. Schwarzschild. Über das Gravitationsfeld eines Massenpunktes nach der Einsteinschen Theorie. *Sitzungsber. Preuss. Akad. d. Wiss.*, pages 189–196, 1916. *http://arxiv.org/abs/physics/9905030*.

[27] R.D. Sorkin. Spacetime and causal sets. *Relativity and gravitation: Classical and quantum*, pages 150–173, 1990.

[28] R.D. Sorkin. Causal sets: Discrete gravity. In *Lectures on quantum gravity*, pages 305–327. Springer, 2005.

[29] O. C. Stoica. Analytic Reissner-Nordström singularity. Phys. Scr., 85(5):055004, 2012.

[30] O. C. Stoica. Beyond the Friedmann-Lemaître-Robertson-Walker Big Bang singularity. *Commun. Theor. Phys.*, 58(4):613–616, 2012.

[31] O. C. Stoica. Schwarzschild singularity is semi-regularizable. Eur. Phys. J. Plus, 127(83):1–8, 2012.

[32] O. C. Stoica. On the Weyl curvature hypothesis. *Ann. of Phys.*, 338:186–194, 2013.

[33] O. C. Stoica. *Singular General Relativity – Ph.D. Thesis.* Minkowski Institute Press, 2013. *http://arxiv.org/abs/1301.2231.*

[34] O. C. Stoica. Cartan's structural equations for degenerate metric. *http://www.mathem.pub.ro/bjga/v19n2/B19-2.htm* Balkan J. Geom. Appl., 19(2):118–126, 2014.

[35] O. C. Stoica. Einstein equation at singularities. *Cent. Eur. J. Phys,* 12:123–131, 2014.

[36] O. C. Stoica. Metric dimensional reduction at singularities with implications to quantum gravity. *Ann. of Phys.*, 347(C):74–91, 2014.

[37] O. C. Stoica. On singular semi-Riemannian manifolds. *Int. J. Geom. Methods Mod. Phys.*, 11(5):1450041, 2014.

[38] O. C. Stoica. The Geometry of Black Hole Singularities. *Advances in High Energy Physics*, 2014:14, 2014. em http://www.hindawi.com/journals/ahep/2014/907518/.

[39] O. C. Stoica. Causal structure and spacetime singularities. *http://arxiv.org/abs/1504.07110*, 2015.

[40] O. C. Stoica. Kerr-Newman solutions with analytic singularity and no closed timelike curves. *U.P.B. Sci Bull. Series A*, 77, 2015.

[41] O. C. Stoica. The Friedmann-Lemaître-Robertson-Walker big bang singularities are well behaved. *Int. J. Theor. Phys.*, 55(1):71–80, 2016.

[42] O. C. Stoica. Spacetime causal structure and dimension from horismotic relation. *Journal of Gravity*, 2016(6151726):1–6, 2016.

[43] O. C. Stoica. The geometry of warped product singularities. *Int. J. Geom. Methods Mod. Phys.*, 14(2):1750024, 2017.

[44] EP Verlinde. Emergent gravity and the dark universe. *http://arxiv.org/abs/1611.02269*, 2016.

[45] Erik Verlinde. On the origin of gravity and the laws of newton. *Journal of High Energy Physics*, 2011(4):1–27, 2011.

6   Determinism and Indeterminism on Closed Timelike Curves

Ruward A. Mulder and Dennis Dieks

Abstract   Notoriously, the Einstein equations of general relativity have solutions in which closed timelike curves (CTCs) occur. On these curves time loops back onto itself, which has exotic consequences: for example, traveling back into one's own past becomes possible. However, in order to make time travel stories consistent constraints have to be satisfied, which prevents seemingly ordinary and plausible processes from occurring. This, and several other "unphysical" features, have motivated many authors to exclude solutions with CTCs from consideration, e.g. by conjecturing a chronology protection law.

In this contribution we shall investigate the nature of one particular class of exotic consequences of CTCs, namely those involving unexpected cases of indeterminism or determinism. *Indeterminism* arises even against the backdrop of the usual deterministic physical theories when CTCs do not cross spacelike hypersurfaces outside of a limited CTC-region—such hypersurfaces fail to be Cauchy surfaces. We shall compare this *CTC-indeterminism* with four other types of indeterminism that have been discussed in the philosophy of physics literature: quantum indeterminism, the indeterminism of the hole argument, non-uniqueness of solutions of differential equations (as in Norton's dome) and lack of predictability due to insufficient data. By contrast, a certain kind of *determinism* appears to arise when an indeterministic theory is applied on a CTC: things cannot be different from what they already were. Again we shall make comparisons, this time with other cases of determination in physics.

We shall argue that on further consideration both this indeterminism and determinism on CTCs turn out to possess analogues in other, familiar areas of physics. CTC-indeterminism is close to the epistemological indeterminism we know from statistical physics, while the "fixedness" typical of CTC-determinism is pervasive in physics. CTC-determinism and CTC-indeterminism therefore do not provide incontrovertible grounds for rejecting CTCs as conceptually inadmissible.

A. S. Stefanov, M. Giovanelli (Eds), *General Relativity 1916 - 2016. Selected peer-reviewed papers presented at the Fourth International Conference on the Nature and Ontology of Spacetime, dedicated to the 100th anniversary of the publication of General Relativity, 30 May - 2 June 2016, Golden Sands, Varna, Bulgaria* (Minkowski Institute Press, Montreal 2017). ISBN 978-1-927763-46-9 (softcover), ISBN 978-1-927763-47-6 (ebook).

1   Introduction

There have been extensive discussions in the philosophical and physical literature of the last couple of decades about the possibilities of time travel: the existence of solutions of the Einstein equations in which closed time-like curves (CTCs) occur has endowed the science-fictional character of the subject with a certain amount of scientific respectability. That the Einstein equations of general relativity do not exclude the existence of CTCs is easy to see. The Einstein equations impose *local* conditions on spacetime: the local curvature properties must stand in a definite relation to the local energy and momentum of the matter fields. As long as these local conditions remain satisfied, the global topology of the spacetime may vary. Now, one particular solution of the Einstein equations is Minkowski spacetime, in which the curvature vanishes everywhere (Minkowski spacetime is flat) and in which there is no matter (all components of the energy-momentum tensor are zero). From this Minkowski spacetime we can build a new solution of the Einstein equations, with a different topology, by the simple operation of identifying two spacelike hypersurfaces (one "in the future", and one "in the past"). Concretely, we cut a strip out of Minkowski spacetime and glue the upper and lower ends together. The cylindrical spacetime that results (a strip of Minkowski spacetime rolled up in the time direction) features CTCs: timelike worldlines going straight up in the time direction return to their exact starting points.

It is helpful to keep this simple example of CTCs provided by rolled-up Minkowski spacetime in mind during the discussion of strange features of CTCs below. However, it should not be thought that CTCs only arise by (arguably artificial) cut-and-paste constructions: there are quite a number of solutions of the Einstein equations known in which apparently plausible matter distributions give rise to CTCs. The most famous solution with CTCs was found by Kurt Gödel (Gödel, 1949, [14]). The Gödel solution describes a non-expanding, rotating universe with a large cosmological constant. This Gödel world is conceptually important, even though it has properties that conflict with what we empirically find in our universe. A solution of the Einstein equations that may be more relevant for our own world is that of a part of spacetime with a spinning black hole (described by the Kerr metric), in which a region with CTCs exists, the *ergosphere* (see, for example, Carroll, 2014, [2], p. 261.). There are also other, less drastic mass-energy distributions that feature CTCs, such as the first metric with CTC-properties to be discovered: van Stockum's rotating cylinder of dust particles (van Stockum, 1937, [24]).

It should be added that it is unclear whether the existence of these and similar solutions has the implication that there are possibilities for building a "time machine". The construction of a manageable time machine would require the creation of a singularity-free compact spacetime region in which CTCs occur, and results by Hawking and others indicate that this cannot be realized within classical general relativity because of energy-conditions that have to be satisfied by the matter fields (Hawking, 1992, [15]). In a quantum theory of general relativity there might be more room for time machines, but this remains speculative in view of the absence of a full quantum gravity theory. However, we are not concerned with the question whether it is possible to construct a useful time machine, but rather with the conceptual status *per se* of CTCs, in relativistic worlds in general.

It is immediately clear that CTCs give rise to causal oddities. If someone could go back into his own past, even if not in our universe but in another physically possible world, it seems that he could change things there in such a way that logical inconsistencies result. The notorious example is the grandfather paradox: if a time traveler arrives at a point in his past at which his grandfather is still an infant, he could decide to kill the child. But if this were to be successful, it would obviously conflict with the very fact of the time traveler's own existence. So in order to have a consistent history on a CTC, certain consistency conditions have to be fulfilled. These conditions take the form of restrictions on what can happen. Although in the case of the grandfather paradox it is plausible at first sight that the time traveler can do whatever is within his capabilities when meeting his grandfather, he nevertheless must be constrained in his actions. This points into the direction of a *determinism* that is stricter than what we are used to in physics; this is one of the points to be further discussed below.

In contrast, there are also cases in which the presence of CTCs appears to lead to a weakening of usual notions of determinism. Think of a space-time that is globally Minkowski-like, but in which there is a finite region containing CTCs that are causally closed within themselves, so that nothing happens on them which has a cause external to the CTC. Closed worldlines of inertially moving particles would be an example. Since there is no causal relation to anything outside such worldlines, the processes that take place in the CTC region cannot be predicted from outside that region. This lack of predictability points into the direction of *indeterminism*, the other issue to be discussed in more detail below.

The threats posed by logical inconsistency, causal anomalies and other strange features have sometimes been adduced to declare the possibility of CTCs "unphysical". CTCs should perhaps be ruled out by a "cosmic censorship principle" (Penrose, 1968, [20]) or a "chronology protection principle" (Hawking, 1992, [15]). The motivation behind conjecturing that such a principle is at work is that the alternative, with its CTC extravagances, goes against the very nature of physics: such features occur nowhere else in existing physical theory or practice. This then is taken to justify the inductive conclusion that they cannot happen at all.

In this paper we shall critically analyze this argument for the impossibility of cases of "unphysical determinism and indeterminism." As we shall argue, it is not accurate to say that the "causal anomalies" associated with CTCs form a category of their own. Indeed, there exist similar cases in standard physics, and these cases are not considered to be exotic or even strange. Therefore, we shall conclude, such *a priori* objections are not insuperable and hence at least part of the motivation for the dismissal of CTCs falls away.

2 Closed Timelike Curves and Logical Consistency

2.1 A Toy Model: Deutsch-Politzer Spacetime

A spacetime that is a bit more complicated than the rolled-up version of Minkowski spacetime that we considered in the Introduction, and which is suitable for illustrating our arguments, is the so-called Deutsch-Politzer spacetime (Deutsch, 1991, [5]; Politzer, 1992, [21]). It can be realized by making two cuts in flat Minkowski spacetime, as indicated in Figure 4—the points in these cuts are removed from the manifold. The two inner edges of the cuts are subsequently identified (glued together); the same happens with the two outer edges. When a particle hits L^- from below it will travel onward from the upper side of L^+. The other way around, a particle hitting the lower side of L^+ will reappear at the upper side of L^-. This creates a region where CTCs can occur: vertical lines, which represent particles at rest (in the depicted frame of reference) that would loop back onto themselves, as depicted by the particle worldline in the figure.

This cut-and-paste operation results in Minkowski spacetime with a kind of "handle", the latter having the internal topological properties of a rolled-up strip of Minkowski spacetime. The spacetime region of the handle can be entered from the rest of spacetime, namely from the two sides, on the left and on the right, where the handle merges with the surrounding spacetime. One can think of this spacetime as global Minkowski spacetime, which at two singular points makes contact with a rolled-up strip of Minkowski spacetime.[1]

Because worldlines inside the handle do not cross the spacelike hypersurface Σ (see Figure 4), Σ is not a Cauchy surface: specification of the physical state on Σ does not fix the physical state on the whole manifold. In particular, there is no information available on Σ about what happens inside the handle. There is consequently a "Cauchy horizon", which limits the part of spacetime that can be predicted from Σ. The unpredictable part comprises the handle, and also the two future lightcones with their apexes in the two singular endpoints of the handle. In Figure 4 these two singularities are represented by the two ends (on the left and right, respectively) of the lower cut. Since the singularities are not part of the manifold, they cannot contain initial conditions for worldlines that "come out of them" (this is a figurative way of speaking, since the singular points are not in the manifold and therefore not on any curve in the manifold—the past parts of the worldlines in question asymptotically approach one of the singularities). These

---

[1]It has been shown that for the Deutsch-Politzer spacetime it is not possible to smooth out the metric such that we would obtain a global nonsingular asympotically flat Lorentzian metric (Chamblin et al., 1994, [3]). This is because the end points of the "cuts" are singularities: as judged from the surrounding Minkowski spacetime, there is a finite spacetime interval between the lower and upper end points on both sides, but seen from the inside these points are identical. These two singularities raise questions about the empirical plausibility of Deutsch-Politzer spacetime. Nevertheless, consideration of this spacetime is helpful as it makes visualization of finite CTC-regions possible; our conclusions will not depend on a commitment to this or another specific spacetime.

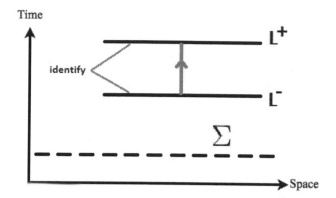

Figure 4: Deutsch-Politzer spacetime, picture is based on (Arntzenius & Maudlin, 2013). L^+ and L^- represent the 'cuts' in Minkowski spacetime; the lower side of L^+ is identified with the upper side of L^-, while the lower side of L^- is identified with the upper side of L^+. This can be visualized as a "handle" on Minkowski spacetime, which particles may enter and leave again (in contrast with the rolled-up cylinder). Σ is a spacelike hypersurface outside of the CTC-region; it is not a Cauchy surface. The green line represents a particle at rest, with a closed timelike worldline.

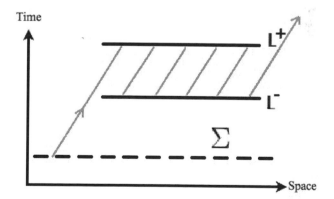

Figure 5: A particle traveling through Deutsch-Politzer spacetime. The particle enters the CTC-region with a certain velocity, and after hitting L^+ five times, it continues its path in the outer Minkowskian region. Picture is based on (Arntzenius and Maudlin, 2013).

worldlines do not have a starting point, and can in this respect be compared to worldlines coming in from infinity. There is no origin of such worldlines whose properties determine their nature or number. Clearly therefore, data available on Σ cannot fix what is beyond the Cauchy horizon.

2.2 Consistency Constraints: The Grandfather 'Paradox'

In the region of the handle in Deutsch-Politzer spacetime CTCs occur. These CTCs raise questions about the consistency of histories and the determination of events, as already mentioned in the Introduction.

For example, a person whose worldline is one of these CTCs will return into his own past, where he must find himself in exactly the same state as on earlier occasions. This uniqueness is simply a logical consequence of the demand that histories on the CTC must be unambiguous: there has to be exactly one physical state of affairs at each point of the CTC. This uniqueness gives rise to a consistency constraint: We can only have solutions that are consistent in the sense that they consist of well-defined unambiguous events along the CTC. This is a logical truism and as such a harmless requirement. Nevertheless, in the present context these consistency constraints introduce a restriction on possible histories that seems counter-intuitive and leads to what Smeenk and Wütrich (2011, [22], p. 7) call *modal paradoxes*, of which the grandfather paradox is a concrete instance.

In the grandfather paradox our time traveler, living on a CTC, goes back in time and meets his grandfather. Obviously, everything will have to happen in exactly the same way as recorded in history. In particular, it cannot happen, on pain of logical inconsistency, that the grandson undertakes actions that make his own birth impossible. The paradox, as usually formulated, is that this seems to take away some of the powers of the grandson: surely, one is apt to argue, he is *able* to kill his grandfather. So how could it be that he cannot in fact do so? Stephen Hawking has argued that such paradoxes threaten time travel with "great logical problems" and that we should hope for "a Chronology Protection Law, to prevent people going back, and killing our parents" (Hawking, 1999, [16]). Such a novel Law is not needed, though: as we shall argue the trivial constraint that everything is well defined and consistent will do the job.

Before discussing this further, we should mention that a more complicated solution of the paradox was suggested by David Deutsch in his quantum model of time travel. Here, physical systems can traverse so-called Deutsch-CTCs that go from one world to another in Everettian many-worlds quantum mechanics (Deutsch and Lockwood, 1994, [6]). This model connects events in different worlds, and the model therefore basically invokes multiple time dimensions—as noted, e.g., in (Dunlap, 2016, [9]). It is true that multidimensional time offers a way out of the grandfather paradox, since the fact that a grandson killed his grandfather in a world with time t_2 does not contradict that his grandfather stays alive in a world with time t_1, in which the grandson was born (cf. Dainton, 2010, [4], p. 123). However, this response of invoking multiple time dimensions seems artificial in solving issues with time travel and, moreover, fails to address the original paradox, in which the time traveler goes back to his own past in his own time in his own world. It is only this original paradox that we shall consider here.

An important remark concerning the original paradox, which takes away some of the puzzlement, was made by David Lewis. As Lewis points out, in the formulation of the paradox there is an ambiguity in the use of 'can' and 'being able' (Lewis, 1976, [17]). Of course, the time traveling grandson *is able* to kill his grandfather in the sense that he knows how to use a firearm, has the required muscular strength, training, and so on. But not everything that 'can' be done in this sense will *actually* be done—we do not at all need to consider time travel to recognize this and to realize that there is no contradiction here. In fact, it is a general truth, also in universes without CTCs, that only one act among all the acts that one is able to do will actually be done. In the case of a CTC, the grandson can accordingly be assumed to be able to kill his grandfather, in the sense of possessing the required means and capacities: he 'can' shoot. But at the same time it is impossible that he will actually do so: if he did, he would contradict the historical record. Accordingly, no immediate contradiction between 'can' and 'cannot' arises. According to Lewis the paradox is therefore only apparent; there is an ambiguity in the word 'can', which is not always visible, but which is highlighted in some cases—and it is highlighted very prominently in time travel situations. That the time travel story does not sit well with common sense is simply because we are not used to CTC-like situations.

Lewis' analysis is correct in our opinion, but its emphasis on human acts and capabilities invites questions, e.g. about the nature of volition and human powers, that distract from the physical aspects of the problem. We shall therefore discuss variations on the paradox that only involve physics. In section 3 we shall consider particles that obey *deterministic* dynamical laws; in section 5 we shall consider an *indeterministic* physical process (like radio-active decay, governed by quantum mechanics). As it will turn out, application of a deterministic theory to a spacetime in which there is a CTC-region leads to a particular kind of indeterminism. In section 4, we shall compare this CTC-indeterminism to other forms of indeterminism we know from physics. By contrast, in the case of an indeterministic quantum process, grandfather-paradox-like reasoning on a CTC will lead us to determinism—at first sight in conflict with what the standard interpretation of quantum mechanics tells us about the essentially indeterministic nature of the theory.

The latter result, the appearance of determinism in an initially indeterministic context, may perhaps be expected since we have already seen that consistency on CTCs reduces possibilities. That there is also a counterpart to this, namely the appearance of *indeterminism*, has also already been indicated, in the previous subsection: beyond the Cauchy horizon there are worldlines that cannot be fully determined from Σ. This can be used to construct examples of indeterminism, even if the local physical laws are deterministic.

3 From Determinism to Indeterminism

In our construction we shall use the Deutsch-Politzer spacetime explained in section 2.1. In Figure 4, the chronology violating region (i.e., the region where CTCs occur) is located to the future of a spacelike hyperplane Σ. No worldlines from this region cross Σ, and in general no wordlines cross Σ more than once. Everything looks therefore "normal" on Σ, just as on an arbitrary hyperplane in Minkowski spacetime. However, the initial value problem on Σ is not well posed because Σ does not qualify as a Cauchy surface. Indeed, the standard definition of a (global) Cauchy surface Σ in a spacetime manifold $\mathcal{M}$ is a surface that is intersected exactly once by *every* non-spacelike curve in $\mathcal{M}$. It is understandable that initial conditions on such a surface Σ determine all events in $\mathcal{M}$ if the applicable laws of physics are locally deterministic, since the physical state on Σ is propagated by these laws along the non-spacelike curves of $\mathcal{M}$. In the case of a Cauchy surface these curves fill the entire spacetime. But in the case of Figure 4 there clearly are worldlines that do not intersect Σ and about whose behavior no information is available on Σ. Consequently, the state on Σ does not contain enough information to fix the entire global state of $\mathcal{M}$.

One could say that the CTC-region to some extent forms a world in itself. It is true that particles can enter the CTC-region from outside, such as in Figure 5, and it is true that this can be predicted from the initial conditions on Σ. However, one may add an arbitrary number of undetermined worldlines with a particle on it, beyond the Cauchy horizon. This will of course lead to different global states of $\mathcal{M}$, with a different total mass and energy. Therefore, associated with any initial state on Σ is an infinitude of global states of $\mathcal{M}$. This seems a clear case of indeterminism, which will typically occur in spacetimes in which isolated chronology violating regions occur. We shall christen this kind of indeterminism "CTC-indeterminism."

4 CTC-Indeterminism Among Other Varieties of Indeterminism

In this section we shall compare the indeterminism that we have seen to arise when CTCs are present with other cases of indeterminism in physics. If the argument that CTCs can be dismissed because of their exotic and unphysical features is to work, the indeterminism that arises here should be in a category of its own, different from cases we encounter elsewhere in physics. In order to see whether this is in fact so, we shall successively review the indeterminism of Norton's Dome, the indeterminism of quantum mechanics, that of the hole argument, and finally the indeterminism of statistical physics.

4.1 Norton's Dome

Norton's Dome is an example in which Newton's second law of motion fails to have one unique solution (Norton, 2008, [18]). In this case the mathematical condition for the relevant differential equation to have a unique solution, the so-called Lipshitz continuity condition, is not satisfied.[2]

The set-up is as follows. Consider a dome-like surface, as shown in Figure 6, in a uniform gravitational field and with a particle of mass m that can move on it. The height $h(r)$ of the surface of the dome as a function of the radial coordinate r, is given by

$$h(r) = \frac{2\alpha}{3g} r^{3/2}, \tag{1}$$

where g is the gravitational acceleration, α a proportionality constant.

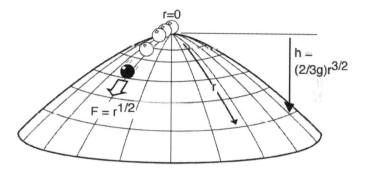

Figure 6: Norton's Dome from (Norton, 2008, [18]). The marble at the top of this dome is initially resting, but will spontaneously—that is, unpredictably—roll down the surface, as shown by the solution 4.

The net force on a particle at the surface will be tangentially directed and is hence given by

$$F = mg \sin \phi = mg \frac{dh}{dr} = \alpha\, mr^{1/2}, \tag{2}$$

with ϕ the angle between the tangent and the horizontal direction. Newton's second law takes the form of the differential equation

$$\frac{d^2 r(t)}{dt^2} = \alpha\, r^{1/2}. \tag{3}$$

If the initial situation is a particle at rest at the top of the dome an obvious solution of Eq. 3 is $r = 0$, $\forall t$. However, there are also other solutions with

---

[2]The Lipshitz condition is the demand that the slope of the force function does not become too large. Specifically, a function F satisfies the condition within a certain domain D iff there is a constant $K > 0$ such that $|F(x) - F(y)| \leq K|x - y|$. For a detailed discussion, see (Fletcher, 2012, [13]).

the same initial condition, namely[3]

$$r(t) = \begin{cases} 0 & \text{if } t < T, \\ \frac{\alpha^2}{144}[(t-T)]^4 & \text{if } t > T, \end{cases} \qquad (4)$$

for an arbitrary value of T. Because of the arbitrariness of T, there is an infinity of possible solutions to this differential equation. Eq. 4 describes a particle initially at the origin, which starts rolling off the surface after the time T has elapsed. Because the initial conditions do not fix one unique motion, determinism fails.

When comparing this indeterminism to our case of CTC-indeterminism, we see that there are various differences. In the dome case, all possibly relevant initial data have been specified, but the differential equation is not able to produce one unique solution from them because the Lipshitz condition is violated. In the CTC case the initial conditions and forces on Σ do produce unique worldlines—Lipshitz conditions are everywhere assumed to be satisfied. But these uniquely determined worldlines departing from Σ do not determine what goes on in the chronology violating region. To know how many particles find themselves in the handle of Figure 4 we need *more initial data*, and these data are not available on Σ. We can conclude that CTC-indeterminism is not connected to the violation of a Lipshitz condition. CTC-indeterminism is therefore essentially different from the indeterminism in the dome case.

4.2   Quantum Indeterminism

The indeterminism of quantum mechanics is given a precise form by the Born rule, which states that the square of the amplitude of a particular term in the quantum state (written down as a superposition in some basis) yields the *probability* for finding a measurement outcome corresponding to that term. According to the standard interpretation of quantum mechanics this probability is fundamental: it is not possible to refine the description by adding parameters to the wave function, in such a way that the predictions become more precise than allowed by the Born rule. In other words, even if the state of a physical system is completely specified, the theory only provides us with *probabilities* for a *range* of possible measurement outcome— the theory is thus indeterministic.

CTC-indeterminism and the indeterminism resulting from the Born rule have in common that in both cases a unique prediction of measurable quantities cannot be fixed by specifying all initial conditions at a given time (that is, say, on some spatial hypersurface). However, in the case of CTC-indeterminism the applicable equations do not tell us anything at all about

---

[3]This is a solution since

$$\frac{d^2}{dt^2}\frac{\alpha^2}{144}(t-T)^4 = \frac{\alpha^2}{12}(t-T)^2 = \alpha\sqrt{\frac{\alpha^2}{144}[(t-T)^4}} = \alpha r^{1/2},$$

satisfying Eq. 3.

probabilities for the different possibilities. One may draw an arbitrary amount of additional causally closed particle worldlines inside the CTC-region, which corresponds to an infinitude of possibilities, but the equations do not assign any chances to them—the equations do not speak about probabilities at all. This is an essential difference with quantum theory, in which the laws possess a probabilistic interpretation from the outset. Put differently, CTC-indeterminism does not come from the probabilistic character of the applicable theory, whereas quantum indeterminism does.

In section 5 we will come back to the status of probabilistic theories when applied to CTC-regions. For now, it suffices to observe that CTC-indeterminism is not similar to quantum indeterminism. In the latter the specification of probability values is essential, whereas in the former probabilities never enter the discussion.

4.3   The Hole Argument

The hole argument, basically devised by Einstein in 1913, makes use of the background independence of general relativity, which ensures that all different coordinate systems perform *a priori* equally well when used to express the laws of the theory—there is no *a priori* given spacetime geometry which could define a privileged frame of reference and coordinates adapted to it (Dieks, 2006, [7]). The modern form of the argument was developed by (Stachel, 2014, [23]) and (Norton and Earman, 1987, [10]), and extensively used in discussions about substantivalism versus relationism with respect to spacetime.

We are concerned with only one possible implication of the hole argument, namely that it proves general relativity to be indeterministic. The argument is as follows. The spacetime of general relativity in a concrete situation contains a labeled set of events placed in a manifold, plus a metric field specifying temporal and spatial relations between the events. Not all the degrees of freedom in the theory are physical, some of them are concerned with how the events are placed in the manifold. Because of the background independence this manifold does not possess a pre-given geometry, so that one can perform a gauge transformation (an active coordinate transformation) on this distribution of events—see Figure 7. All observable properties can be reduced to combinations of relativistic invariants and are left intact by such a transformation. If the transformation constitutes a real change in the world (as the spacetime substantivalist would maintain) then this real change—a different distribution of matter and energy over the manifold—cannot be dealt with by the theory: the initial and boundary conditions do not fix which one of the different possible distributions will be realized. This is because any two distributions of metric and matter agree on all observational properties, which are the only properties predicted by the dynamical laws of the theory. This amounts to indeterminism.

Compared to CTC-indeterminism, this type of indeterminism shares the characteristic that it leaves open an infinite range of possibilities, and does not assign any probabilities to these different options. Moreover, like CTC-

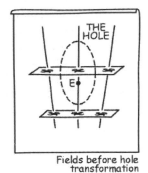 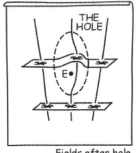

| Fields before hole transformation | Fields after hole transformation |

Figure 7: The hole argument as presented by Norton (Norton, 2015, [19]). The invariant aspects of metric and matter fields of particles (here, even entire galaxies) and their distribution in spacetime are the same, although the events are differently placed in the manifold. The transformation from one situation to the other is accomplished by a 'hole transformation' (a diffeomorphism) in the area indicated by the dotted line. The question then is: does the galaxy pass through spacetime point E or not?

indeterminism, hole indeterminism does not depend on violations of Lipshitz conditions: all differential equations are perfectly well behaved and have unique solutions in terms of invariant quantities. It is only because of general covariance, relating to the absence of a fixed spacetime background, that equally valid spacetime *representations* of the physical situation (characterized by invariant quantities) arise—these representations relate to each other via diffeomorphisms.

The latter observation marks an essential difference between CTC-indeterminism and hole indeterminism. In the case of hole indeterminism all diffeomorphically related possibilities feature exactly the same values of all physical quantities like energy, mass, etc. For this reason, it may be argued that these different solutions are actually physically identical—as relationists do. This is completely unlike the situation in CTC-indeterminism: here the possibilities are uncontroversially physically different, distinguishable as they are on the basis of the numbers of particles inside the CTC-region, the amounts of mass and energy, and so on. So CTC-indeterminism and hole indeterminism belong to very different categories.

4.4 Lack of Knowledge

When the state of a physical system is incompletely specified, it is to be expected that its future behavior cannot be fully predicted, even if the applicable laws are deterministic and if the Lipshitz conditions are satisfied. For example, if we only know the initial positions and velocities of a restricted number of particles, or only possess estimates for these quantities, Newton's equations will not enable us to accurately predict the future state of a many-particle system. More than one final state will be compatible with the initial data.

An example of such a situation in physical practice is the micro-description of systems characterized by macro-quantities, as in statistical physics. Statistical physics applies when the number of degrees of freedom of a system becomes too large to be practically tractable, as in the case of a macroscopic amount of a gas or the modeling of noise-effects in electronic devices. Although the laws of classical mechanics fix a unique evolution given the initial state, it is practically impossible to ascertain all individual initial momentum and position values. Statistical physics deals with this by considering ensembles of *possible* microstates.

The crucial point is that the macrostate underdetermines the microstate: there are many microstates that would give rise to the same macrostate. Hence, statistical physics can be seen as quantifying our ignorance about the microphysical state. This leads us to a "pragmatic" type of indeterminism: in the absence of a full set of initial conditions, the future lies open—in the epistemic sense.

Clearly, this underdetermination is only to be expected when information is missing. The associated indeterminism is consequently harmless from a fundamental point of view: determinism can be restored by making the description more complete. In the ontological sense, only one microstate is actually realized, and this state has a fully deterministic evolution. In the next section, we argue that this non-fundamental indeterminism is very similar to CTC-indeterminism.

4.5 CTC-Indeterminism as Epistemic and Harmless

CTC-indeterminism comes about as the result of the existence of Cauchy horizons, which restricts the amount of information available on spacelike hypersurfaces—what is beyond the horizon is hidden from view. This indicates that CTC-indeterminism is of the same kind as the epistemically founded indeterminism just discussed, and hence of the harmless kind. This suggestion is supported by our earlier observation that CTC-indeterminism is fundamentally different from more problematic kinds of indeterminism in physics (Norton's Dome, quantum indeterminism and diffeomorphism-indeterminism).

Indeed, there is a direct analogy between CTC-indeterminism and the indeterminism arising from lack of knowledge as in statistical physics. In statistical physics data about the macrophysics do not fix the initial conditions of microscopic particles; in the CTC case, initial conditions are hidden from view by Cauchy horizons. Although in statistical physics the microstate is underdetermined, this does not mean that there is no unique future of the system ontologically speaking: according to classical mechanics there is one unique set of initial conditions, one evolution and hence no indeterminism. In the case of CTC-indeterminism, there is also a fact of the matter concerning initial conditions and the precise shapes of the worldlines behind the horizon. Just as we can take away the lack of knowledge and the associated indeterminism in statistical physics by more fully specifying initial conditions, CTC-indeterminism can be taken away by specifying conditions beyond the horizon, namely at the upper side of L^- of Figure 4 and at arbitrarily small spheres surrounding the two singular points (points on the lightcones emerging from these points in Figure 4). These conditions added to the initial conditions on Σ fully complement the initial-value problem and hence restore global determinism.

Our intuition might suggest that the specification of additional initial conditions on a hypersurface outside Σ is unnatural. However, there is nothing in classical mechanics or relativity theory that implies that Cauchy surfaces should have the form of global spatial hypersurfaces. Even in classical theory one can easily define situations in which conditions on a plane cannot fix everything that is going to happen in the future. Think, for example, of a box with a region in it that is shielded from electromagnetic fields. This is a topological structure that in relevant aspects is analogous to that of Figure 4: initial data inside the shielded region will be necessary to achieve a globally deterministic description. Similarly, taking away indeterminism by specifying initial conditions in the region behind the Cauchy horizon is the natural thing to do in time travel situations.

That there may be a need to explicitly look at the perhaps unusual nature of the global situation in order to determine initial conditions may seem counter-intuitive. However, this feeling is likely due to our impression that we directly experience the world at large, and have access to infinite space-like hypersurfaces—from which we can make predictions. In reality, however, already special relativity teaches us that we can only have knowledge about limited spacetime regions so that our actual predictions have a local character. In other words, we do not know from experience that there are global Cauchy hypersurfaces and should be open to the possibility that our world is different. From this perspective, CTC-indeterminism is not stranger and more worrisome than the indeterminism arising from lack of knowledge in other areas of physics.

5 From Indeterminism to Determinism

There is an interesting counterpart to the indeterminism that arises in the presence of CTCs. As we already noted, consistency of histories on CTCs requires that constraints are satisfied that guarantee the uniqueness of events along each CTC. In the case of, e.g., the grandfather paradox these constraints lead to restrictions on what a human agent can do when returning to his own past—this suggests a kind of "superdeterminism" that imposes stronger conditions than what we are inclined to expect. The "lack of freedom" that results from this is a well-known reason not to take CTCs seriously.

As noted in section 2.2, appeals to human agency and free will can easily obscure the physical points that are at stake. Fot this reason we shall analyze a variation on the grandfather paradox in which a quantum process is considered instead of the actions of a human time traveler. Think of an electron in a quantum state that is a superposition of two different eigenstates of energy, both with equal weights, and with energies E_1 and E_2, respectively. After some time the electron arrives at a measuring device designed to measure energy and interacts with it. Quantum theory predicts that there is a 50%

probability that E_1 will be found and recorded, and an equal probability that E_2 will be recorded.[4]

According to the standard interpretation of quantum mechanics there are no underlying deterministic factors that determine the outcome of this experiment. The quantum state is taken to contain a full specification of everything that is relevant to the prediction, and this state only yields probabilities. In fact, one can derive technical results (Bell's inequalities and their violation) showing that the addition of "hidden variables" to the quantum state necessarily leads to theoretical schemes in which the added electron properties behave in unusual and undesirable ways: the hidden properties cannot be purely local (i.e., they must exert an instantaneous influence on each other, even when belonging to physical systems that are far apart). So the indeterminism of quantum mechanics is different from a simple consequence of a lack of knowledge of causal factors. If a proof could be given that quantum indeterminism nevertheless cannot be fundamental, this would be a far-reaching result.

Yet, the consistency conditions that are in force on CTCs seem to imply precisely such a non-fundamentality of quantum mechanical probabilities. Indeed, when we imagine the above-described experiment on a CTC, exactly *one* of the two possible outcomes will be realized—let us assume it is E_1. Now, in the future of this outcome we return to the same quantum state of the electron that existed at the beginning of our experiment. Everything will "repeat itself", so that the outcome of the experiment will necessarily be E_1, the same as before. So there is no question of any probability or uncertainty: the outcome is completely fixed; this is logically forced due to the consistency constraints. Since there is only one unique outcome event on the CTC, with unambiguous properties, we have to conclude that this outcome is determined *tout court*: what is recorded at the end of the experiment is an intrinsically fixed and determined event and there is no place for indeterminism.

If this argument is correct, the feeling that CTCs should be banned from physics by the introduction of a "chronology protection principle" may well be justified. Indeterminism as a possible fundamental feature of physical reality would be disproved by a simple CTC thought experiment. This seems to fly in the face of an enormous amount of foundational work in quantum theory. So this version of the grandfather paradox, even though not threatening inconsistency, seems worrisome—at least at first sight.

6 Determinism versus Determination

The argument that CTCs are "unphysical" because of CTC-determinism, like the earlier argument from indeterminism, relies on the supposition that we

---

[4]The frequently discussed case of a superposition of *spin* states, e.g. the singlet state, is basically identical. It is important to note that the predictions for the outcomes of measurements made by quantum mechanics do not depend on the presence of human observers; quantum predictions are statements about what is recorded by macroscopic measuring devices, regardless of whether a conscious agent becomes aware of the outcomes.

are facing a consequence that only arises in the presence of CTCs. We have argued that this strategy does not work in the case of CTC-*indeterminism*, because this indeterminism also occurs in other situations and is considered harmless there. We will argue now that the case of CTC-*determinism* does not fare better. As a first step we shall show that in many cases CTC-determinism merely exemplifies a kind of logical *determination* that trivially characterizes physical processes in general and does not conflict with *physical* indeterminism.

To see the difference between physical *determinism* and *determination*, think of Minkowski spacetime (Newtonian spacetime will do as well) with a fundamentally indeterministic physical theory defined on it. Physical indeterminism implies that the complete physical state at a certain instant does not completely fix, via the laws of the theory, the physical state at later instants: more than one later states are compatible with the initial state according to the theory in question. Nevertheless, the later states are *determined* in a trivial, logical sense: they are exactly what they are and nothing else. This is just the requirement of unambiguity of events that is needed to make sense of the idea of four-dimensional spacetime at all. If there were no unique physical state of affairs at each spacetime point, we could not have a well-defined history of the universe.

The indispensability of this kind of unambiguous *determination* is especially clear within the conceptual framework of the so-called block universe, associated with the "B-theory" of time. According to this B-theory there are no absolute ontological distinctions between Past, Present and Future, and the whole of history can be thought of as laid out in one four-dimensional "block". Evidently, all events in this block must be *determined* in the sense just mentioned. It is true that in the literature this has sometimes been taken to entail that the block universe necessarily is subject to determinism. However, this is now generally recognized as a fallacy, at least if *physical determinism* is meant (cf. Dieks, 2014, [8]). Indeed, the block universe exists, according to the B-theory of time, independently of whether the events in it are generated by deterministic laws, of a theory like classical mechanics, or by indeterministic laws like those of quantum mechanics. The crux is that the distinction between physical determinism and physical indeterminism is a distinction between two different kinds of *relations* between physical states: in the deterministic case the laws of the theory make exactly *one* later state compatible with the earlier one, in the indeterministic case there are more later states compatible with the earlier one, according to the theory. By contrast, in the case of logical determination the question of what physical theory applies is not relevant: for an event to be determined in this logical sense it is sufficient that the event is well defined, regardless of its relations to other events. Any *bona fide* unambiguous event is *determined* in this sense.

This argument may seem to rely on the B-theory of time and the associated ontology of the block universe, but on closer inspection it does not. Even if one subscribes to the "A-theory of time", and accordingly believes that there are ontological differences between Past, Present and Future, it remains tautological that each future event *will be* exactly and uniquely what it *will be*, and that past events *were* exactly what they *were*. This tautology merely requires that there is exactly *one* history of the world. So even here, all events are determined in the logical sense, regardless of whether a deterministic or an indeterministic physical theory applies.

Now think back of our thought experiment involving an electron on a CTC. We assumed that indeterministic quantum theory governed what happened in the experiment. This means that the physical state of our electron before its interaction with the measuring device is compatible with different measurement outcomes. This is an assumption about the *relation* between two different states, defined at two different temporal stages of the experiment: the laws of quantum mechanics do not completely fix the measurement outcome when given the initial electron state as their input. This relation between the initial and final physical states is objective and unambiguously defined, even on a CTC: when in thought we follow the electron in its history, we return to exactly the same states as before when we arrive at the same points in the electron's existence. So the relation between any two states is uniquely defined. This relation can certainly be the indeterministic one specified by quantum mechanics. Nevertheless, the final outcome of the experiment is unambiguously determined since it corresponds to the single and unique measurement event (that we encounter again and again when we let our mind's eye traverse the CTC various times).

This is fully analogous to what takes place in situations *without* CTCs. It is perhaps easiest to recognize this by again thinking of the block universe: according to quantum theory there are probabilistic relations between the physical properties instantiated at different spacelike hypersurfaces in this universe, but nevertheless these physical properties are unambiguously and uniquely determined in themselves. As we have already mentioned, the same argument can be used even if the notion of a block universe is rejected and some version of the A-theory of time is adopted. Accordingly, if we accepted the principle to brand as *unphysical* processes that are governed by indeterministic physical theories but are nevertheless fully *determined* in the logical sense, we would have to throw out all of physics.

However, this is not the whole story. The plausibility of the argument that CTCs lead to (super)determinism also derives from the fact that *information* about the final outcome may already be present when the experiment starts. For example, we might imagine that the outcome is recorded in a book, which survives along the CTC and can be consulted at the beginning of the experiment. In this case it seems natural to assume that the physical state at the moment that the experiment starts contains this information, so that it becomes possible to derive the experiment's outcome from this initial state. This then appears to lead to physical determinism after all.

In order to discuss this argument and its relevance for the acceptability of CTCs we have to delve a bit deeper into the question of what the physical *laws* on the CTCs look like, and whether these laws enable us to predict the outcome of experiments from their initial states as just indicated.

7 Prediction and Retrodiction on CTCs

Suppose that the quantum experiment that we described in the previous section, with possible outcomes E_1 and E_2, is performed on a CTC and has the actual outcome E_1. Suppose further that a record of this is preserved along the CTC, up to the point at which the experiment is about to start. The total physical state at this point on the CTC does not only comprise the state of the electron, but also the record that reveals the outcome of the experiment. Therefore, in spite of the fact that the quantum state of the electron can only provide us with probabilities, the total state makes it possible to predict with certainty what the outcome of the experiment will be.

It should be noted that for this "deterministic" prediction the quantum state is not needed at all: the record does not supply a piece of missing information that is needed to supplement what is given by the quantum state, but does all the predictive work on its own. Logically speaking, this is not different from the laboratory situation (without any CTCs) in which a quantum experiment is done, its outcome noted down, after which the record is put in an archive from which we can retrodict the outcome even many years later. This retrodiction does not rely on a quantum mechanical calculation using the later state of the electron (if it still exists), but is based on the classical behavior of records. Books and similar records are designed to be more or less permanent and to follow the deterministic laws of classical physics to a very high degree of approximation. If we only had the later quantum state of the electron to guide us, we would not be able to retrodict unequivocally what the experiment's outcome was; what makes the retrodiction reliable is the approximately classical behavior of the record. The situation in our thought experiment on the CTC is very similar. It is not the case that quantum theory has ceased to be applicable, but an additional factor has come into play, namely the presence of a classical deterministic record. When the experimenter consults the book, at the beginning of the experiment, and notes the recorded outcome, she is engaged in exactly the same activity as an observer who consults his notebook in our Minkowski-like world to see what has happened. So just as we are not entitled to conclude that reliable retrodiction falsifies the fundamentally probabilistic nature of quantum mechanics in our world, can we draw that conclusion in the case of the CTC thought experiment.

One might object that the CTC experiment is essentially different from the familiar retrodiction case: what is at stake is not retrodiction at all but rather *prediction* of the outcome of an experiment, on the basis of the initial physical state. But this counterargument does not succeed: the reliability of the statement that the outcome of the experiment will be as the book indicates, derives solely from the book's reliability as a retrodictor, in the same way as in the case without CTCs. The future-directed aspect only comes in because of the uniqueness (and determination) of events along the CTC: as it happens, the retrodicted event is the very same event as the future result of the experiment. So it is the combination of retrodiction and determination that is at work here, and not some new physical principle of prediction.

This becomes more transparent when we ask whether there are reasons to assume the existence of new lawlike physical principles on the CTC. For example, will there be a law telling us about the future? This should not be expected. It is not a general feature of CTCs that there are notebooks or similar records of the eventual outcome available at the beginning of each experiment—these may well have been erased or eroded long ago, if they ever existed at all. There is no reason to assume new regularities on CTCs that ensure that records will be more stable than usual; quite the opposite, if the outcome is to be recorded at the end of an experiment, any already existing record will have to be erased before. It is perfectly consistent to assume the usual laws of physics on CTCs plus the boundary condition of periodicity, i.e. the consistency constaints, without worrying about any new laws.

The consistency conditions that apply in worlds with CTCs thus need not be regarded as symptoms of the existence of novel physical principles, as has sometimes been suggested in the literature.[5] They just reflect uniqueness of events and, indeed, consistency; they have a logical rather than a physical character. There are no fixed patterns of events connected to them. The most one can say is that these consistency conditions enforce a violation of a principle that one is inclined to employ in calculations in which no CTCs are involved, namely that arbitrary initial conditions can be imposed locally— one only needs to think of rolled-up Minkowski spacetime to see that not all such specifications will lead to the periodical solutions that we need to be consistent.

The idea (which we argue is violated in the CTC-case) that arbitrary initial conditions can be imposed locally has been formulated by Deutsch and Lockwood (Deutsch & Lockwood, 1994, [6], p. 71), which they have called the Autonomy Principle:

> it is possible to create in our immediate environment any config-
> uration of matter that the laws of physics permit locally, without
> reference to what the rest of the universe may be doing.

It is true that *we are not used to constraints on local initial conditions*—but then again, we are not used to taking into account global considerations at all.

---

[5]For example, Earman (Earman, 1995, [11], p. 194) writes: "Indeed, the existence of consistency constraints is a strong hint—but nevertheless a hint that most of the literature on time travel has managed to ignore—that it is naive to expect that the laws of a time travel world which is nomologically accessible from our would will be identical with the laws of our world. In some time travel worlds it is plausible that the MRL laws [Mill, Ramsey and Lewis, or the "best systems account"] include the consistency constraints; in these cases the grandfather paradox has a satisfying resolution. In other cases the status of the consistency constraints remains obscure; in these cases the grandfather paradox leaves a residual itch. Those who wish to scratch the itch further may want to explore other analyses of laws. Indeed, time travel would seem to provide a good testing ground for competing analyses of laws." In our view, the consistency constraints merely reflect consistency, which of course has to be satisfied as a trivial logical principle anyway, and do not introduce new laws of physics.

But, in principle, this is mistaken even in worlds without any CTCs, as we shall note in the Conclusion. At the end of the day, the constraints in question are nothing but the expression of ordinary physical laws in combination with the principles of logic and do not require new principles of physics.

8 Conclusion

Common sense may rule out such excursions—but the laws of physics do not.
—David Deutsch & Michael Lockwood (1994, p. 69).

The potential existence of closed timelike curves poses a challenge to our intuitions concerning determinism and indeterminism. Physical processes inside a time travel region must satisfy consistency conditions, which constrain the dynamics, and on the other hand the existence of Cauchy horizons implies a lack of predictability. However, when judging the seriousness of this conflict with intuition we should not forget that our common-sense notions about (in)determinism and predictability derive from untutored interpretations of everyday experience.

One such untutored common-sense idea is that we have immediate access to a global now, a plane, from which the world develops to its future states. Relativity theory has taught us that this notion is mistaken: we do not have epistemic access to a global now at all because of the existence of a finite maximum speed of signal propagation, and ontologically the theory does not single out preferred now-planes. In keeping with this, we should adapt our ideas about causality from global to local notions. It is a basic message of relativity theory that criteria for determinism should first of all depend on local considerations, and not on considerations about the possible existence of global nows in the universe.

CTC-indeterminism hence turns out to be similar to a familiar and harmless type of indeterminism that occurs in many other places in physics. It is very different from the varieties of philosophically interesting indeterminism that have been focused on in the recent literature, and rather represents another instance of epistemically grounded lack of predictability that is pervasive in physics. From a local point of view, everything is completely deterministic and the lack of global predictability can be rectified by adding extra, not yet considered initial conditions. CTC-indeterminism is therefore not the extraordinary new phenomenon that it sometimes has been suggested to be.

Something similar can be said about the determinism in the sense of "lack of freedom" that results from consistency constraints on CTCs. What we are facing here is basically determination of events at the level of logic, instead of physical determinism. Also in this case there is no reason to think that new physical principles, too strange to be true, have to be invoked.

Summing up, both CTC-indeterminism and CTC-determinism do not involve new principles or new laws of physics; although intuitively strange at first sight, analysis shows that they are of the same kind as cases already familiar from well-known and accepted applications of physical theory. As a consequence, there is no justification for the argument that we are here facing phenomena that are so exotic that their potential presence suffices to rule out CTCs. What the *prima facie* implausibility of CTC-determinism and CTC-indeterminism shows is that we have not yet succeeded in adapting our intuitions to the world of general relativity—and that we are unexperienced with traveling through time.

References

[1] Arntzenius, F., T. Maudlin (2013). "Time Travel and Modern Physics." *The Stanford Encyclopedia of Philosophy*. Editor: E. N. Zalta. URL: http://plato.stanford.edu/archives/win2013/entries/time-travel-phys/.

[2] Carroll, S. (2014). *Spacetime and Geometry: An Introduction to General Relativity*. Harlow: Pearson Education Limited.

[3] Chamblin, A., G. W. Gibbons, A. R. Steif (1994). "Kinks and Time Machines." *Physical Review D*, 50, R2353–5.

[4] Dainton, B. (2010). *Time and Space*. Oxon; New York: Routledge, 2nd Edition.

[5] Deutsch, D. (1991). "Quantum mechanics near closed timelike lines." *Physical Review D*, 44, 3197–3217.

[6] Deutsch, D., M. Lockwood (1994). "The Quantum Physics of Time Travel." *Scientific American*, March 1994, 69–74.

[7] Dieks, D. (2006). "Another look at general covariance and the equivalence of reference frames." *Studies in History and Philosophy of Modern Physics*, 37, 174–191.

[8] Dieks, D. (2014). "Time in Special Relativity." In *Springer Handbook of Time*, 91–113. Editors: A. Ashtekar and V. Petkov. Berlin-Heidelberg: Springer.

[9] Dunlap, L. (2016). "The Metaphysics of D-CTCs: On the Underlying Assumptions of Deutsch's Quantum Solution to the Paradoxes of Time Travel." *History and Philosophy of Modern Physics*, 56, 39–47.

[10] Earman J., J. D. Norton (1987). "What Price Spacetime Substantivalism." *British Journal for the Philosophy of Science*, 38, 515–525.

[11] Earman, J. (1995). *Bangs, Crunches, Whimpers, and Shrieks. Singularities and Acausalities in Relativistic Spacetimes*. New York; Oxford: Oxford University Press.

[12] Earman, J. (2004). "Determinism: What We Have Learned and What We Still Don't Know." In *Feedom and Determinism*, 21–46. Editors: J. K. Campbell, M. O'Rourke, D. Shier. Cambridge, MA and London: The MIT Press.

[13] Fletcher, S. (2012). "What Counts as a Newtonian System?: The View from Norton's Dome." *European Journal for Philosophy of Science*, 2, 275–297.

[14] Gödel, K. (1949). "An Example of a New Type of Cosmological Solutions of Einstein's Field Equations of Gravitation." *Reviews of Modern Physics*, 21, 447–450.

[15] Hawking, S.W. (1992). "Chronology protection conjecture." *Physical Review D*, 46, 603–611.

[16] Hawking, S. W. (1999) "Space and Time Warps," online lecture, URL: `http://www.hawking.org.uk/space-and-time-warps.html`.

[17] Lewis, D. (1976). "The Paradoxes of Time Travel." *American Philosophical Quarterly*, 13, 145–152.

[18] Norton, J. D. (2008). "The Dome: An Unexpectedly Simple Failure of Determinism." *Philosophy of Science*, 75, 786–798.

[19] Norton, J. D. (2015). "The Hole Argument." *The Stanford Encyclopedia of Philosophy*. Editor: E. N. Zalta. URL: `http://plato.stanford.edu/archives/fall2015/entries/spacetime-holearg/`.

[20] Penrose, R. (1969). "Gravitational Collapse: The Role of General Relativity." *Rivista del Nuovo Cimento*, Serie I, Numero Speziale I, 252–276.

[21] Politzer, H. D. (1992). "Simple quantum systems in spacetimes with closed timelike curves." *Physical Review D*, 46, 4470–4476.

[22] Smeenk, C., C. Wütrich (2011). "Time Travel and Time Machines." In: *Oxford Handbook of Time*, 577–630. Editor: Craig Callender. Oxford University Press.

[23] Stachel, J. (2014). "The Hole Argument and Some Physical and Philosophical Implications." *Living Reviews in Relativity*, 17: 1. doi:10.12942/lrr-2014-1.

[24] Stockum, W. J. van (1937). "The gravitational field of a distribution of particles rotating around an axis of symmetry." *Proceedings of the Royal Society, Edinburgh*, 57, 135–154.

Part II

SPACETIME AND QUANTUM THEORY

A. S. Stefanov, M. Giovanelli (Eds), *General Relativity 1916 - 2016. Selected peer-reviewed papers presented at the Fourth International Conference on the Nature and Ontology of Spacetime, dedicated to the 100th anniversary of the publication of General Relativity, 30 May - 2 June 2016, Golden Sands, Varna, Bulgaria* (Minkowski Institute Press, Montreal 2017). ISBN 978-1-927763-46-9 (softcover), ISBN 978-1-927763-47-6 (ebook).

7   Duality and Emergence

Dennis Dieks

Abstract   During the last decades cases of "duality" have become a focus of attention in theoretical physics. Dual theories are theories that look very different, with dissimilar mathematical structures, but predict identical values of physical quantities. An important example is provided by theories that bear a "holographic" relation to each other: here a gravitational theory defined within a spatial volume gives the same empirical results as a theory *without* gravitation that is defined on the surface of that volume. This is not infrequently taken to suggest that the volume (or "bulk") theory, with its added gravity and extra dimension, in some way is "generated" by the lower-dimensional surface theory—put differently, that gravity and the new spatial dimension *emerge* from a more fundamental layer of reality that is gravitationless and non-spatial. In this paper we shall critically investigate the idea that duality can in this way be responsible for emergence. After an explanation of the notion of emergence itself, as it occurs in physics, we shall argue that in relevant cases exact dualities should be seen as situations in which we are facing different formulations of precisely the same theory rather than as examples of emergence. This raises the question of what is needed for real cases of emergence. In order to answer this question we shall briefly discuss a recent proposal that treats gravity as truly emergent.

1   Introduction: Emergence and Duality

Cases of emergence, in which novel phenomena arise from an underlying layer of reality that itself is characterized by very different features, are well known in physics. A familiar example is the emergence of the macroscopic world, which "surfaces from" what happens at the level of atomic and subatomic particles and their interactions. Related to this, the paradigm case of emergence in physical *theory* is the appearance of thermodynamics as a result of averaging over processes involving atoms and molecules. In the emergent theory (thermodynamics) concepts like temperature and pressure take center stage—concepts that are not even defined on the more fundamental level of atoms and molecules. These new concepts figure in macroscopic laws that

A. S. Stefanov, M. Giovanelli (Eds), *General Relativity 1916 - 2016. Selected peer-reviewed papers presented at the Fourth International Conference on the Nature and Ontology of Spacetime, dedicated to the 100th anniversary of the publication of General Relativity, 30 May - 2 June 2016, Golden Sands, Varna, Bulgaria* (Minkowski Institute Press, Montreal 2017). ISBN 978-1-927763-46-9 (softcover), ISBN 978-1-927763-47-6 (ebook).

are emergent as well.

It is a typical feature of this and similar cases, and as Butterfield (2011) argues of the notion of emergence in general, that novel, stable and characteristic patterns of behavior arise at a level of description at which underlying degrees of freedom become relatively unimportant and can be averaged over (as happens when we go from the level of atomic and molecular processes to the level of macroscopic thermal phenomena). The emergent behavior is consequently relatively insensitive to the details of what is going on at the level of the more basic processes—as is exemplified by the general features of thermal macroscopic behavior. Indeed, the general properties of phase transitions and of equations of state are independent of the specific details of microscopic interactions.

This example will be our cue for how to use the term "emergence": emergent features play a central role in stable regularities that are characteristic of a new level of description. They can be explained—in principle—from what goes on at a more basic level of description, but usually it is unfeasible in practice to use this more fundamental approach directly. Importantly, on the level of the emergent phenomena a "bird's eye view" can be adopted that is relatively independent of details on the fundamental level.

In this paper we are interested in the relation between emergence, in the sense just discussed, and *duality*, a feature of theories that has recently attracted a lot of attention, in particular in the context of quantum gravity research. Duality is a one-to-one mapping between the physical quantities predicted by two physical theories.[1] Such dualities do not need to transform complete formalisms into one another in a one-to one way; indeed, the mathematical form of the two theories may be very different, for example because they use spacetime backgrounds of unequal numbers of dimensions. So it does not need to be the case that states are transformed into states and Hamiltonians into Hamiltonians, as usual in symmetry transformations; and the two theories may use entirely different recipes for forming physical quantities. The only thing of importance for the presence of duality is that two theories agree about the numerical values of the predicted physical quantities.

A possible case of physical emergence that has received much attention in the recent physics literature concerns the possibly non-spatiotemporal and non-gravitational background of gravitational theories (general relativity and theories of quantum gravity—though the latter of course have been at best only partially established). That gravity might originate from some deeper layer of reality and in this respect may differ from other forces has a certain intuitive appeal. Indeed, gravity distinguishes itself because it is universal: it applies to all forms of matter and energy in the same way—this reminds one of the universality of thermodynamic descriptions, which as we have seen is a consequence of insensitivity to microscopic details. Moreover, gravity is

---

[1]We are focussing on *exact* dualities here, according to which identical values of physical quantities occur on both sides of the mapping. The case of approximate dualities is conceptually different, and this is relevant for the question of whether the concept of emergence is applicable, as we shall discuss below.

notoriously more difficult to quantize than other forces. This might reflect a difference of principle from other physical forces occurring in present day theoretical physics, e.g. in the standard model.

In addition, as we shall see in a moment, recent research in black hole physics has led to the hypothesis that the physical predictions of quantum gravity theories within a spatial volume correspond to the predictions made by theories *without gravitation* that are defined on the *boundary* of this volume. This provides us with a real-life example of a duality from present-day frontier physics. It seems only a small step from this duality between a gravitational and a non-gravitational theory to the view that gravity *emerges* from processes described by a theory without gravity. It is this latter notion that we shall critically analyze here.[2]

2   The Holographic Hypothesis

The central ideas of holography stem from developments in the early nineteen nineties. In particular, in a seminal paper 't Hooft (1993) surmised (see also (Susskind, 1995)) that the number of degrees of freedom of a black hole is proportional to the surface area of the black hole's horizon. This not only tells us how many degrees of freedom there are in a black hole system, but it also suggests that one does not need a theory about the interior of the black hole at all to understand what goes on inside. Indeed, because the number of degrees of freedom contained in the interior is completely represented on its boundary surface area, this number does not scale with the size of the volume, so that the three-dimensional description of the volume does not seem to play a determinative role. As 't Hooft put it, "We suspect that there simply *are* not more degrees of freedom to talk about than the ones one can draw on a surface [...]. The situation can be compared with a hologram of a three dimensional image on a two dimensional surface ('t Hooft, 1993, p. 6)." This statement contains the essence of the holographic hypothesis.

In the actual case of a three-dimensional hologram it is legitimate to see the two-dimensional information carrier as the source that generates the three-dimensional holographic picture, via causal chains of interfering beams of laser light. It is therefore not surprising that 't Hooft stated in the abstract of his 1993 paper that the holographic hypothesis naturally leads one to believe that gravitational processes in three-space are a manifestation of things that really happen in a space with fewer dimensions and without gravity: "our world is not 3+1 dimensional" ('t Hooft, 1993, p.1). There is an explanatory arrow here that goes from surface to bulk, with the plausible implication that the surface theory is more basic than the theory of the enclosed volume. One is certainly tempted to express this by saying that the space-time theory of the enclosed volume *emerges* from the description on the surface.

---

[2]See (Dieks et al., 2015) for more technical details about many of the topics of this article.

However, in other places of his 1993 paper 't Hooft was less explicit about the relation between the surface and the bulk theories; the paper was programmatic and did not contain a worked-out example of two concrete dual theories. But this situation of tentativeness has changed by the vast amount of recent work on the so-called "AdS/CFT duality". This duality provides us with a concrete example of a holographic relation between two theories, and we shall briefly discuss it in the following section.

3 The AdS/CFT Duality

The idea of a holographic correspondence between a gravitational bulk theory and a gravitationless theory defined on the boundary of the bulk spacetime has found an explicit illustration in string theories for quantum gravity in Anti-de Sitter (AdS) spacetime. Anti-de Sitter spacetime is a particular solution of the Einstein equations of general relativity. In non-quantum general relativity the structure of this spacetime fixes not only all temporal and spatial intervals, but also tells us how particles move under the influence of gravity: point particles follow geodesics in the geometrical structure if only gravity is at work. Attempts to develop a quantum version of this gravitational picture include the postulation of "strings" governed by the equations of quantum theory. The hope is that these elementary building blocks and their interactions can be made responsible for a quantum version of gravity. Now, it turns out that there are strong theoretical reasons for believing that such string theories in AdS spacetime correspond exactly to so-called "conformal field theories" (CFT), quantum theories in which there is no gravity, on the *boundary* of the AdS spacetime (which, of course, has one spatial dimension less than Anti-de Sitter spacetime itself). This "AdS/CFT duality" was first conjectured by Juan Maldacena (1997). More generally, the AdS/CFT duality relates string theory in $d + 1$-dimensional Anti-de Sitter spacetime (AdS_{d+1}) to a conformal field theory on a d-dimensional space isomorphic to the boundary of AdS.[3]

The correspondence between the two sides of the duality (string theory in AdS and conformal field theory on the boundary of AdS, respectively) is formally represented by the numerical equality of the "partition functions" on both sides. These partition functions are central theoretical quantities: knowledge of the partition function makes it possible to determine all quantum expectation values of physical quantities, which together constitute the physical content of the theory. It follows that the partition function of string theory in AdS spacetime, which codifies all information about physical quantities in this spacetime, fully determines all physical quantities predicted by the surface theory—and *vice versa*. If the holographic hypothesiss is correct in the AdS/CFT case, the bulk and the surface theories therefore share the same values of physical quantities (numerical values of expectation values).

---

[3]A conformal field theory is a quantum field theory that is invariant under conformal transformations, that is, coordinate transformations that multiply the metric by a scalar function (the "conformal factor").

In discussions of duality, especially in the context of AdS/CFT, it is sometimes stated that one is dealing with two theories that are the "same." For example, in his textbook on string theory Zwiebach (2004, p. 376) describes the situation thus: "the term 'duality' is generally used by physicists to refer to the relationship between two systems that have very different descriptions but identical physics." What such characterizations clearly intend to express is the just-mentioned existence of a one-to-one correspondence between the expectation values of physical quantities (observables) on the two sides of the duality ("observable" is here used in its technical quantum mechanical sense and so refers to physical quantities that could be measured *in principle*, however indirectly; there is no direct relation to observability by the unaided human senses). However, it should not be forgotten that the full theoretical structures of the dual theories look very different. For example, the line element of AdS^4 does not occur at all in the geometrical structure of the dual theory (a conformal field theory on flat Minkowski spacetime), so that there is no isomorphism between the full mathematical structures of the respective theories. It is remarkable that theories that are formally so dissimilar still are able to yield the same numbers. This certainly suggests some physical relation between the two sides of the duality.

One may wonder whether dual theories do not they share more structural properties than the one-to-one mapping between their observables and expectation values. One should expect, for example, that they should share all symmetries between observables. These correspondences between symmetries are indeed found in concrete examples. For instance, the space-time symmetry group of the CFT in d dimensions ($SO(2,d)$) equals the isometry group of $(d+1)$-dimensional AdS; and there are also other matches of symmetries (e.g. supersymmetries).

The situation as sketched raises the question as to the exact nature of the relation between dual theories. The equality of the partition functions makes one suspect that numerically correct accounts of any conceivable experiment or problem phrased with one theory's objects and concepts can be modeled using the concepts and objects of the dual theory. Can therefore a case be made that these theories are really the same in spite of their formal differences, as suggested by Zwiebach and others? Or might it be possible to define an explanatory relation between the members of the dual pair, which could justify considering one of them as less fundamental than the other and perhaps as "emerging"? Or should we rather conclude that we are dealing with coincidental numerical similarities and analogies that do not imply a deep physical relation between the theories?

4 The Interpretation of Duality

The distinction between exact and approximate dualities is relevant when we attempt to answer such questions. In the case of an approximate duality, in

---

$^4 ds^2 = \frac{l^2}{r^2}(dr^2 - dt^2 + d\vec{x}^2).$

which there is no *exact* match between the predictions of the two theories, uncontroversial criteria for theory evaluation and comparison can be deployed. In this situation there may certainly be scope for the notion of emergence: approximate duality is an inter-theoretic relation that could very well be like the relation between thermodynamics and statistical mechanics. Also in the latter case, one description approximates the other; and it is undisputed that the atomistic/molecular description is more fundamental than the macroscopic one. In principle, even though usually not in practice, the microscopic predictions are always better than the thermodynamic ones, and there are cases (for example Brownian motion) in which this is in fact of practical importance. Our earlier characterization of emergence is clearly satisfied in this paradigm case. On the macroscopic, thermodynamic level concepts like "temperature" and "pressure" become applicable and play an essential role in the thermodynamic laws; in other words, these concepts characterize novel and robust behavior that is largely insensitive to the specifics of the underlying atomistic and molecular dynamics. Thermodynamics emerges from the (statistical) mechanics of atoms and molecules, and this involves an asymmetry between the theories: thermodynamics comes forth from the micro description, but not the other way around. A relation of *approximate* duality might be similar concerning these aspects and may in this case give rise to an effective description that emerges from a more fundamental one.

However, the standard assumption in the holography and AdS/CFT literature is that the dualities are *exact*, and here it is less clear how we should judge the relative status of the dual theories. By definition there is in this case a precise one-to-one mapping between values of physical quantities: for each physically significant number in one theory there is an identical counterpart in the other.

This one-to-one mapping suggests complete empirical equivalence of the two theories: the theories appear to predict exactly the same things in whatever situation, so that it is not possible to distinguish between them on purely empirical grounds. However, it should be noted that a one-to-one mapping between values of physical quantities by itself does not imply empirical equivalence. Indeed, the mapping may relate different quantities, with different physical meanings. This situation in fact often occurs in the practice of physics, e.g. when models are made of situations described by one theory with the help of another, perhaps completely different theory. In these situations it is presupposed that we know what the physical meaning of the quantities in each of the theories is. For instance, in actual practice we know what terms such as "electrostatic potential" and "fluid velocity" mean. This makes it possible for us to mimic situations in electrostatics by means of a hydrodynamic model, because the same equations govern the behavior of these different quantities in the respective theories.

This modeling presupposes the existence of what one might call an "external point of view", independent of the two theories, from which the reference relation between each of the theories and physical reality can be fixed. In such cases, duality between theories expresses a *symmetry* in the physical world: exactly the same relations that obtain between, say, the potentials in certain electrostatic processes are also satisfied by fluid velocities in stationary fluid flows.

A well-known example of this kind of duality is provided by the source-free Maxwell equations, which exhibit perfect symmetry between the E and B fields. When we consider the application of these equations to a source-free region of space, the form invariance under an exchange of E and B reflects a physical symmetry that is present in this region. If other regions do contain charges, and if it is supposed that we can identify these charges independently of our choice for one of the two variants of Maxwell's theory, this breaks the symmetry and determines unambiguously what the symbols E and B refer to.

In this case exchanging E and B in the source-free region does not change anything in the local form of the equations, but it does imply a drastic change in the physical situation that is described: electric fields are replaced by magnetic fields. So here the duality connects *different* aspects of the physical world (the electric and magnetic fields in a charge-free region) that possess an isomorphic internal structure. As already pointed out, situations of this kind enable us to make *models* for physical phenomena falling under one theory with the help of concepts from another theory.

Clearly, in these examples of modeling the notion that the duality is connected with emergence does not even suggest itself. With regard to the last example: the symmetry between E and B does not entail anything about a possible origin of electricity in magnetism or the other way around.

The situation becomes more interesting, and more in the spirit of discussions about duality in the context of present-day fundamental physics, in the case of an exact duality between two theories that both aspire to give global descriptions of the same world (including all experiments and their outcomes). In this situation, it is no longer clear that there exists an "external" point of view that independently fixes the meanings of terms in the two theories. To revert to the above electrodynamic example: if two variants of Maxwell's theory were to connect E and B differently to electric charges, and we would not have independent means to identify the relation between charges and fields, it would become impossible to find out whether we find ourselves in a theory described by variant one or variant two. More realistically, think of AdS/CFT: although we use a concept in the CFT that we denote by the term "energy", the very idea of holography is that the associated symbol represents, via the duality, *distance* relations in the bulk theory: there exists a strict correspondence between CFT-*energies* and bulk *distances*. Now, how would it be possible to establish that this symbol really refers to energy rather than distance, or vice versa? If we do not get information from outside of the two theories, from an independent external perspective telling us in which world we live, it is impossible to answer this question. So if the holographic idea in the form originally proposed by 't Hooft (all degrees of freedom in the bulk are represented by degrees of freedom on the surface, and vice versa) really works there is no reason to believe that "energy" in CFT and "distance" in the bulk do not refer to one and the same thing.

Generally speaking, in cases of duality like this one we are dealing with two structures of observables and their values that have exactly the same internal relations to each other in the two respective theories. Without an independent external viewpoint, the only thing to go on with regard to the meaning of these observables is how they are positioned within their two respective networks of relations. But this means that we are justified in concluding that the isomorphism between the structures of observables can be cashed out in terms of *equality* rather than *symmetry*. The symbols used in the two theories may very well be different, but in view of the identical roles quantities play in relation to other quantities, and the sameness of the values they assume, *identifications* can be established: A in one theory will denote exactly the same feature of physical reality as B denotes in the other if A and B occupy structurally identical nodes in their respective webs of observables and assume the same (expectation) values.[5]

In the philosophical literature, it is frequently argued that to identify two empirically equivalent theories boils down to instrumentalism.[6] The thought behind this is that differences in theoretical structure between theories may well correspond to differences in physical reality, even if these differences are not (yet) observable: we should not assume that the descriptive physical content of theories is exhausted by the theory's observable consequences. But in our cases of exact duality the situation is different. As we have pointed out before, the "observables" that are in one-to-one correspondence with each other in cases of exact duality are not defined via a notion of observability as in the debate between empiricists and scientific realists. Rather, they stand for what is physically real and meaningful according to the theories under discussion (i.e. quantum expectation values of all physical quantities), even if there are no possibilities of direct observation. So what we are facing is not the standard situation of empirical equivalence in which two different physical theories coincide "on the surface of observable phenomena": we are dealing with theories that coincide exactly on everything that each of them deems physically real.

We therefore conclude that in the case of an exact duality between theories without fixed external rules of correspondence a very strong form of equivalence arises; but one that does *not* lead to theoretical underdetermination. Since in this case the two theories completely agree on everything that is physically meaningful, the two sides of the duality should be taken as different representations of one and the same physical world. They consequently represent only one theory; and there is no emergence of one side of the duality from the other.

Dualities therefore either connect *different* parts of physical reality that display the same structures, in which case they signal the presence of symmet-

---

[5] What is assumed here is not necessarily a structuralist doctrine about what the world is like, but rather a view about how a mathematically formulated "theory of everything" can be assumed to correspond to the world if no *a priori* "rules of correspondence" between the theory and world are given. In this case it is only the internal structure of the theory that can help to decide how it mirrors the world.

[6] See (Acuña and Dieks, 2014) for a more extensive discussion of empirical equivalence and theoretical underdetermination, and their philosophical implications.

rical features of the world; or they connect different formulations of exactly the same theory. In neither case can it be said that there are features on one side of the duality that emerge from features on the other side (assuming the meaning of emergence explained in section 1). Emergence needs more than empirical equivalence: it needs an asymmetry of the kind that is present in the relation between statistical physics and electrodynamics. A story about gravity as an emergent phenomenon therefore has to contain more than holography and duality. An example of what such a theory of emergent gravity could look like is provided by a recent proposal by Verlinde (2011). In the following we shall sketch this proposal in order to illustrate what it adds to holography.

5   Gravity as an Emergent Phenomenon

In his proposal Verlinde starts from holography. He begins with a submicroscopic theory without gravity on a two-dimensional "screen", e.g. the surface of a sphere, with the intention of arriving at a description of *gravitational* processes in the interior of the sphere. As we now know, this holographic correspondence in itself will not suffice for emergence, so that we need an additional ingredient. What Verlinde adds is that the submicroscopic theory is going to be considered in a thermodynamic limiting situation: it is only in this limit that gravity will arise. In fact, the idea is that gravity will result as a mere "effective force" that is generated by elementary processes in which there are no forces at work at all. Clearly, a scenario like this can be realized in the arena of emergence as we have defined it in section 1.

In more detail, Verlinde invites us to consider the surface of a sphere and to assume that some gravitationless theory governs what happens there. A salient point of his proposal is that it is possible to remain quite unspecific about the precise nature of this submicroscopic theory: it is sufficient to assume that the theory describes physical processes that can be characterized in a very general information-theoretic way, as "changes in information". The surface area of the sphere is therefore imagined to be divided in small cells, each of which may contain one or zero bits of information. A global physical situation corresponds to a distribution of 0s and 1s over the cells, while time evolution of the global state corresponds to a change of this distribution over the cells. The holographic principle makes us expect that this informational process on the surface will encode information about what goes on inside the sphere, for example about where matter is located in the sphere's interior and how masses fall inward.

The number of bits on the sphere is taken to be extremely large, which makes it possible to assume that an effective thermodynamic description can be used. From the viewpoint of thermodynamics the physical process that takes place on the surface (changes in the distribution of the 0s and 1s over the cells) can be characterized as a process that maximizes entropy: the distribution will tend to an equilibrium distribution.

The idea that a thermodynamic account of what happens on the surface may correspond to a description in terms of gravity pertaining to the sphere's interior is now made plausible through a deduction of Newton's law of gravity, via a holographic translation of thermodynamical equations on the surface. Without claiming that this derivation of Newton's law is in all respects convincing or even valid at all, we shall reproduce its general idea here in order to illustrate how in principle gravity might come forth from non-gravitational processes as an emergent phenomenon.

First, it is assumed that there is a number of active bits on the sphere, N, which is proportional to the sphere's total surface area A:

$$N = \frac{Ac^3}{G\hbar} \, , \tag{1}$$

in which the area of the sphere can be written as

$$A = 4\pi R^2. \tag{2}$$

Although the terminology that is used suggests otherwise, no embedding of the surface in a three-dimensional world is presupposed: in the surface theory R should be considered as a quantity that is *defined* by Eqs. (2) and (1). As already stated, it is essential for the derivation that the thermodynamic limit can be taken on the surface. A *temperature* T will therefore be definable, and it is assumed that in thermodynamic equilibrium this temperature relates to the total energy E on the sphere through the law of equipartition:

$$E = \frac{1}{2}Nk_BT \, . \tag{3}$$

We can now *define* the quantity M by:

$$E = Mc^2. \tag{4}$$

On the surface, M is just an alternative expression for the thermodynamic energy; but via holographic correspondence it will soon acquire the interpretation of the total gravitational mass that is present in the interior.

Now a change in the entropy on the surface that corresponds to what happens if, via the holographic translation, a particle with mass m changes its position by Δx is assumed to be[7]:

$$\Delta S = 2\pi k_B \frac{mc}{\hbar} \Delta x \, . \tag{5}$$

In the thermodynamic description on the screen the redistribution of informational bits can be treated as a process towards equilibrium, in which the entropy grows. Such processes can be characterized phenomenologically as the result of the operation of an effective "entropic force" F that represents

---

[7]This follows Bekenstein's 1973 formula; see (Dieks et al., 2015) for more details on the justification of this assumption.

the effects of changes in entropy:

$$F \Delta x = T \Delta S .$$ (6)

The peculiarity of an entropic force is that it does not derive from an interaction, but arises from the statistical behavior of large numbers of microscopic degrees of freedom. A typical example is the force that can be used to describe the behavior of a polymer, stretched in the direction Δx. On the fundamental level, viz. the level of the atoms that make up the polymer, everything happens without forces: the polymer may consist of short chains of atoms that are connected but can rotate freely with respect to each other. However, as a result of random microscopic motion, the polymer will with overwhelming probability end up in a macroscopic state that corresponds to a large phase space volume; this will be a state in which the polymer is coiled up (the central idea is that there are vastly more coiled-up microstates than states in which the chains of the polymer are collinear). So from the macroscopic point of view a definite directedness in the behaviour of the polymer manifests itself: it tends to coil up, as a result of mere microscopic randomness. This tendency (associated with an increase in entropy) can be phenomenologically described as caused by an elastic force obeying Eq. (6). A more garden-variety example would be the dispersion of an ink droplet in a swimming pool: although this is the effect of random motion of molecules, at the macroscopic level it could appear as if a force drove the droplet apart.

Going back to our case, we have to remember that the growth of entropy on the sphere is holographically related to the motion of masses in the bulk. We can now determine the magnitude of the effective force associated with this motion, by simply combining the above relations (1–6). This yields the result

$$F = G \frac{Mm}{R^2} .$$ (7)

This, surprisingly, is Newton's law of gravitation. Its derivation in the way sketched suggests that gravity is an *entropic force* whose "corresponding potential has no microscopic meaning", as Verlinde (2011, p. 4) comments.

In his paper Verlinde introduces a variation on essentially the same idea in order to generalize his derivation into one that yields the Einstein equations of general relativity. A related derivation of the Einstein equations was given earlier by T. Jacobson (1995), who already claimed that these equations are "born in the thermodynamic limit" (p. 1260). Also other authors have proposed that gravity should be seen as an emergent phenomenon in this sense; see e.g. (Barcelo et al., 2001), (Konopka et al., 2008), (Hu, 2009), (Oriti, 2014), (Padmarabhan, 2015). Verlinde's proposal distinguishes itself by the explicit and intuitively clear interpretation of gravity as an entropic force (Verlinde, 2011, p. 9), and the discussion of this in the context of holography and duality.

As is clear from the above sketch of the derivation, two assumptions in it are essential for the appearance of gravity from the gravitationless surface theory, namely 1) the holographic correspondence between surface and bulk, and 2) the transition from the microscopic to the thermodynamic mode of description, which grounds the characterization of gravity as an entropic phenomenon. The holographic correspondence here consists in a mere *translation* between surface and bulk concepts; it does not contribute to the emergence of gravity (in the sense of emergence of section 1). That gravity is an *emergent* phenomenon in Verlinde's proposal is solely due to the thermodynamic limit that is taken, in accordance with our analysis in the previous section.

6 Conclusion: Emergence and Duality

We have considered three scenarios: 't Hooft's original holographic proposal, AdS/CFT, and Verlinde's recent scheme. Although in 't Hooft's 1993 introduction of the holographic hypothesis it was suggested in places that the physics on the boundary is more fundamental and explanatory than the physics in the bulk, there were no sustained claims of emergence.'t Hooft was more concerned with the question of how many degrees of freedom are needed to describe the interior of a black hole than with questions of ontological or epistemic priority.

However, in the AdS/CFT case, conceived as an example of exact duality, many have expressed the feeling that the bulk and its gravitational description can be seen as consequences of what happens on the surrounding surface; that the bulk description and gravity *emerge* from the boundary field theory.[8] It should have become clear from what we have argued that this is not a tenable viewpoint. In the case of an exact duality there is no room for emergence: either the two dual theories refer to different aspects of reality, or—more plausibly in the situation of the universal theories we are dealing with—they represent different formulations of one physical description and should be regarded as one and the same theory. In both cases there is no reason to think that one theory is more fundamental than the other. There is no emergence in the sense of section 1, but rather mutual translatability.

By contrast, in the Verlinden scheme and in similar thermodynamic proposals made by other authors gravity really *is* emergent. This becomes particularly clear when we think of the analogy between gravity and elastic forces that have an entropic character. These entropic forces are effective in the sense that they come from an underlying level in which microscopic counterparts to them do not exist. They appear solely as results of the growth of entropy.

If gravity is really emergent in this sense, this implies that there is no point in looking for a fundamental quantum theory of gravity, or in seeking gravity's unification with the forces of the standard model. Obviously, this constitutes a very important change in perspective: traditionally, the most important problem of quantum gravity is the difficulty to quantize gravity along the lines that have proved successful in quantizing other forces. Furthermore, if gravity is a thermal phenomenon, one may expect that the equations describing it (like Newton's law) represent the behavior of "macroscopic averages" and that

---

[8] See for examples of this interpretation of duality (Dieks et al., 2015).

tiny deviations will occur from what these equations predict—analogous to Brownian motion deviations from the second law of thermodynamics. The conceptual analysis of duality and emergence thus possesses direct relevance for theoretical and experimental research at the frontiers of physics.

References

Acuña, P. and Dieks, D. (2014). "Another look at empirical equivalence and underdetermination of theory choice," *European Journal for Philosophy of Science*, 4, 153–180.

Barcelo, C., Visser, M., Liberati, S. (2001). "Einstein gravity as an emergent phenomenon?", *International Journal of Modern Physics D* 10, 799–806.

Bekenstein, J. (1973). "Black holes and entropy," *Physical Review D*, 7, 2333–2346.

Butterfield, J. (2011). "Emergence, reduction and supervenience: a varied landscape," *Foundations of Physics*, 41, 920–959.

Dieks, D., van Dongen, J. and de Haro, S. (2015). "Emergence in holographic scenarios for gravity," *Studies in History and Philosophy of Modern Physics*, 52, 203–216.

Hooft, G. 't. (1993). "Dimensional reduction in quantum gravity," in: Ali, A., J. Ellis and S. Randjbar-Daemi, *Salam Festschrift*. Singapore: World Scientific.

Hu, B. L. (2009). "Emergent/quantum gravity: Macro/micro structures of spacetime," *Journal of Physics Conference Series* 174, 012015.

Jacobson, T. (1995). "Thermodynamics of spacetime: The Einstein equation of state," *Physical Review Letters*, 87, 1260–1263.

Konopka, T., Markopoulou, F., Severini, S. (2008). "Quantum Graphity: A Model of emergent locality," *Physical Review D* 77, 104029.

Maldacena, J. (1998). "The large N limit of superconformal field theories and supergravity," *Advances in Theoretical and Mathematical Physics* 2, 231–252.

Oriti, D. (2014). "Disappearance and emergence of space and time in quantum gravity," *Studies in History and Philosophy of Modern Physics*, 46, 186–199.

Padmanabhan, T. (2015). "Emergent gravity paradigm: Recent progress," *Modern Physics Letters A* 30 (03n04), 1540007.

Susskind, L. (1995). "The world as a hologram," *Journal of Mathematical Physics*, 36, 6377–6396.

Verlinde, E. (2011). "On the origin of gravity and the laws of Newton," *Journal of High Energy Physics*, 2011 029.

Zwiebach, B. (2004). *A first course in string theory.* Cambridge: Cambridge University Press.

8 Quantum Walk on Spin Network and the Golden Ratio as the Fundamental Constant of Nature

Klee Irwin, Marcelo M. Amaral,
Raymond Aschheim, Fang Fang

Abstract   We apply a discrete quantum walk from a quantum particle on a discrete quantum spacetime from loop quantum gravity and show that the related entanglement entropy drives an entropic force. We apply these concepts in a model where walker positions are topologically encoded on a spin network.

Then, we discuss the role of the golden ratio in fundamental physics by addressing charge and length quantization and by analyzing the ratios of fundamental constants—the limits of nature. The limit of minimal length and volume arising in quantum gravity theory indicates an underlying principle that we develop herein.

1 Introduction

One of the principal results from Loop Quantum Gravity (LQG) is a discrete spacetime—a network of loops implemented by spin networks [1] acting as the digital/computational substrate of reality. In order to better understand this substrate, it is natural to use tools from quantum information / quantum computation. Gravity, from a general perspective, has been studied with thermodynamic methods. In recent years, numerous questions on black hole entropy and entanglement entropy have made this an active field of research. In terms of quantum information and quantum computation, advances have been achieved with the aid of many new mathematical tools. Herein, we present the development of one such tool, which we call the discrete-time quantum walk (DQW). We will see that the problem of a quantum particle on a fixed spin network background from LQG can be worked out with the DQW. This gives rise to a new understanding of entanglement entropy and

A. S. Stefanov, M. Giovanelli (Eds), *General Relativity 1916 - 2016. Selected peer-reviewed papers presented at the Fourth International Conference on the Nature and Ontology of Spacetime, dedicated to the 100th anniversary of the publication of General Relativity, 30 May - 2 June 2016, Golden Sands, Varna, Bulgaria* (Minkowski Institute Press, Montreal 2017). ISBN 978-1-927763-46-9 (softcover), ISBN 978-1-927763-47-6 (ebook).

entropic force, permitting the proposal of a model for dynamics. In terms of physical ontology, we suggest dynamics and mass emerge from this spin network topology, as implemented by the DQW.

In summary, the first part of this paper is a reinterpretation of results from LQG that emphasizes the quantum information perspective of a quantum-geometric spacetime. That is, it adopts Wheeler's *it from bit* and the newer *it from qubit* ontologies—the general digital physics viewpoint.

One important view of reality is the digital physics paradigm—the idea that reality is numerical at its core [2, 3, 4, 5, 6, 7, 8, 9, 10, 11, 12, 13]. In the second part of this paper, we defend the conjecture that reality is information-theoretic at its core by presenting experimental and mathematical justification of our conjecture that the golden ratio is the fundamental dimensionless constant of nature. Numbers include shape-symbols, such as simplex-integers [14]. Generally, a symbol is an object that represents itself or another object. Symbols can be self-referential and participate in self-referential codes or languages, such as quasicrystals [14, 2, 15]. Generally, symbols are highly subjective, where meaning is at the whim of language users. However, simplexes,[1] as geometric number and set theoretic symbols which represent themselves, have virtually no subjectivity. That is, their numeric, set theoretic and geometric meaning is implied via geometric first principles. If space, time and particles were patterns in a quantized geometric code, low subjectivity geometric symbols, such as simplex-integers, could serve as both spatiotemporal and numerical quanta of space while also being part of an organized code or language called a quasicrystalline phason code. One element of digital physics we discuss here is the discreteness of space-time and charge in general—something that a geometric code is well suited to provide.

The *voxel* of space is an old idea that, today, has acquired a body of experimental evidence as well as mathematical proofs [16]. Its early construction starts with the birth of quantum field theory, where it was noticed that models become more regular with a cutoff. When one considers quantum gravity, like Bronstein did in 1936 [17], this question becomes fundamental because, as opposed to smooth spacetime field theories, gravity does not allow an arbitrarily high concentration of charge in a small region of spacetime. Bronstein concluded that this leads to an inevitable limit of the precision to which one can measure the strength of the gravitational field. Advances on different fronts in the following years lead to a general understanding that this *voxel* of space, a minimal volume, must be the foundational building block of any realistic proposal of quantization of the gravitational field. Such deductions provide support for the digital physics view. Quantum mechanics indicates that nature must have a minimum length, volume and time. We can cite the rigorous developments of string theory, LQG, the Maldacena conjecture, black hole physics, various thought experiments, the generalized uncertainty principle and many other works as support of the discretized

---

[1] An n-simplex is an n-dimensional polytope which is the convex hull of its $n+1$ vertices. For example the triangle in 2D, tetrahedron in 3D, 4-simplex in 4D...

spacetime view. For a review, references and historical details, we recommend reference [16]. The discrete nature of quantum gravity theories implies that time and motion are ordered numerical sets on point-like substructures, which can be viewed as discrete pre-spacetime graphs acting as possibility spaces for graph actions. Upon such graph drawings (a geometric representation of a graph), a quantum gravity code theoretic formalism can express itself non-deterministically. This approach leads to a new ontology for the substrate of reality that may replace older ontologies based on either randomness or determinism. The new ontology is called the code theoretic axiom [2], where a finite set of irreducible geometric symbols and a finite set of non-deterministic syntactical rules express as code legal and physically realistic manifestations. Self-organizing expressions of a quantum code are inherently non-deterministic because codes/languages are non-deterministic.

Herein, we address implications of minimal volume, the quantum of charge of the gravitational field (along with minimal charge in general) in light of the code theoretic view. The experimentally validated result of electric charge quantization has a standard theoretical explanation in Dirac magnetic monopoles. This field, which applies topological arguments to address charge quantization, follows Dirac's work in general [18]. We go further here to understand that there are underlying principles governing charge quantization. This is revealed when one considers the Planck volume limit of space. One consequence we discuss is the special role of the golden ratio, where we show that fundamental limits of nature can be understood in terms of the golden ratio.

2 Methods

2.1 Particle interacting LQG

We start by considering a quantum particle on a quantum gravitational field from LQG [19]. In LQG, spin networks define quantum states of the gravitational field. To consider a quantum particle on this gravitational field, we consider the state space built from tensor product of the gravity state $\mathcal{H}_{LQG}$ and the particle state $\mathcal{H}_P$, $\mathcal{H} = \mathcal{H}_{LQG} \otimes \mathcal{H}_P$. The LQG state space can be spanned by a spin network basis $|s\rangle$ that is a spin network graph drawing $\Gamma = (V(\Gamma), L(\Gamma))$, with $V(\Gamma) \rightarrow v_1, v_2...$, vertices coloring and $L(\Gamma) \rightarrow l_1, l_2...$, ($\{\frac{1}{2}, 1, \frac{3}{2}...\}$) links coloring. For the particle state space the relevant contribution comes from the position on the vertices of graph Γ, spanned by $|x_n\rangle$ where ($n = 1, 2...$). The quantum state of the particle captures information from discrete geometry and cannot be considered independently from it. Therefore the Hilbert space is spanned by $|s, x_n\rangle$ and the Hamiltonian operator can be derived by fixing a spin network to the graph drawing, and calculating the matrix element of this operator $\langle\psi|H|\psi\rangle$ with $\psi \in H_P$. Accordingly, for the interaction between the particle and the fixed gravity state space [19] we have,

$$\langle \psi | H | \psi \rangle = \kappa \sum_l j_l (j_l + 1) \left(\psi(l_f) - \psi(l_i) \right)^2 , \tag{1}$$

where κ is a constant that we can initially take as equal to one, l_f are the final points of the link l and l_i the initial points. The interaction term takes this form because the relevant Hilbert space depends on the wave functions at the vertices. So if the link l starts at vertex m and ends at vertex n we can change the notation, relabeling the color of this link l between m and n as j_{mn} and the wave function on the end points as $\psi(v_m)$, $\psi(v_n)$. Now, for H, we have

$$\langle \psi | H | \psi \rangle = \kappa \sum_l j_{mn} (j_{mn} + 1) \left(\psi(v_n) - \psi(v_m) \right)^2 , \tag{2}$$

The operator H is positive and semi-definite. The ground state of H corresponds to the case where the particle is maximally delocalized. This leads to entropy [19]. In reference [20] it was considered that the classical random walk is associated with (2), a Markov chain. The transition probabilities for this random walk are

$$P_{mn} = \frac{j_{mn}(j_{mn} + 1)}{\sum_k j_{mk}(j_{mk} + 1)} . \tag{3}$$

and [20] shows that this reproduces an entropic force. This random walk is implemented with the Laplacian solution (2) and differs from the Laplacians coming from the discrete calculus calculate in [21]. We will consider DQW relations with more general Laplacians in future work.

2.2   DQW

As study the influence of discrete geometry on a quantum field. Here, it is natural to consider the quantum version of random walks. To have DQWs we need an auxiliary subspace, the coin toss space, to define the unitary evolution

$$U = S(C \otimes I), \tag{4}$$

where S is a swap operation that changes the position to a neighbor node and C is the coin toss operator related to this auxiliary subspace. On some graphs its clear how to define the coin toss Hilbert space [22] but for the spin network considered above its not clear exactly what auxiliary space to use. For a more general approach, we can make use of Szegedy's DQW [23, 24], which we can utilize in two ways: 1- consider a bipartite walk or 2 - a walk with memory. For a bipartite walk, if we consider a graph Γ, we can simply make an operation of duplication to obtain a second graph $\bar{\Gamma}$. For our purpose here, it is better to consider the second option − the walk with memory, considering the N_v-dimensional Hilbert space $\mathcal{H}_n$, $\{|n>, n = 1, 2, ..., N_v\}$ and $\mathcal{H}_m$, $\{|m>, m = 1, 2, ..., N_v\}$, where N_v is the number of the vertex

$V(\Gamma)$. The state of the walk is given as the product $\mathcal{H}_n^{N_v} \otimes \mathcal{H}_m^{N_v}$ spanned by these bases. That is, by states at the previous $|m\rangle$ and current $|n\rangle$ steps, defined by

$$|\psi_n(t)\rangle = \sum_m^{N_v} \sqrt{P_{mn}} \, |n\rangle \otimes |m\rangle \,, \tag{5}$$

where P_{mn} is the transition probabilities that define a classical random walk, a Markov chain, which is a discrete time stochastic process without a memory, with

$$\sum_n P_{mn} = 1. \tag{6}$$

Note that (6) is implied by definition of (3). For the evolution, we can consider a simplified version of Szegedy's DQW [25] by defining a reflection, which we can interpret with the unitary coin toss operator,[2]

$$C = 2\sum_n |\psi_n\rangle \langle \psi_n| - I, \tag{7}$$

and a reflection with inverse action of the P swap (previous and current step) that can be implemented by a generalized swap operation

$$S = \sum_{n,m} |m,n\rangle \langle n,m| \,, \tag{8}$$

where we have the unitary evolution

$$U = CS, \tag{9}$$

that defines the DQW.

So using Szegedy's approach is a straightforward way to obtain the discrete quantum walk on the spin network Γ considered above. It is given by equations (5 - 9) with P given by (3). Namely, for equation (5)

$$|\psi_n(t)\rangle = \sum_m^{N_v} \sqrt{\frac{j_{mn}(j_{mn}+1)}{\sum_k j_{mk}(j_{mk}+1)}} \, |n\rangle \otimes |m\rangle \,. \tag{10}$$

We can interpret the coin toss space as the space of decisions encapsulating a nondeterministic process possessing memory, such that we have a unitary evolution. This approach makes the usual entropy convert into entanglement entropy between steps in time. Szegedy's DQW is a general algorithm quantizing a Markov chain defined by transition probabilities $P_{n,m}$. These transition probabilities are obtained directly from the Hamiltonian of the system such that Szegedy's DQW is used to simulate this system with quantum computation [26].

---

[2]Szegedy's DQW does not generally use a coin toss operator in the literature.

2.3 Entanglement Entropy and Entropic Force

We turn now to calculate entanglement entropy. Consider the Schmidt decomposition. Take a Hilbert space $\mathcal{H}$ and decompose it into two subspaces $\mathcal{H}_1$ of dimension N_1 and $\mathcal{H}_2$ of dimension $N_2 \geq N_1$, so

$$\mathcal{H} = \mathcal{H}_1 \otimes \mathcal{H}_2. \tag{11}$$

Let $|\psi\rangle \in \mathcal{H}_1 \otimes \mathcal{H}_2$, and $\{|\psi_i^1\rangle\} \subset \mathcal{H}_1$, $\{|\psi_i^2\rangle\} \subset \mathcal{H}_2$, and positive real numbers $\{\lambda_i\}$, then the Schmidt decomposition would be

$$|\psi\rangle = \sum_i^{N_1} \sqrt{\lambda_i}\, |\psi_i^1\rangle \otimes |\psi_i^2\rangle, \tag{12}$$

where λ_i are the Schmidt coefficients and the number of the terms in the sum is the Schmidt rank, which we label N. With this, we can calculate the entanglement entropy between the two subspaces

$$S_E(C) = S_E(P) = -\sum_{i \in N} \lambda_i log \lambda_i. \tag{13}$$

We can now calculate the local entanglement entropy between the previous step and the current n step (similar for current and next steps). Identifying the Schmidt coefficients λ_i with P_{mn} given by (3) and, from (10), we see that the Schmidt rank N is the valence rank of the node. Then insert in (12, 13), and the local entanglement entropy on current step is

$$S_{E_n} = -\sum_m^N P_{mn} log P_{mn}. \tag{14}$$

By maximizing Entanglement Entropy

$$S_{E_n} = log N_{max}, \tag{15}$$

where N_{max} is the largest valence. Which gives the entropic force for gravity worked out in [20, 27] with $\frac{dS}{dx} = |S_{E_n} - S_{E_m}|$ proportional to a small number identified with the particle mass M

$$\frac{dS}{dx} = |S_{E_n} - S_{E_m}| = \alpha M, \tag{16}$$

where, if we take the logarithm of (15) to be base 2, α is a constant of dimension $[bit/mass]$.

Each of these interpretations has related applications. For example, with the DQW, we have unitary evolution by encoding a non-deterministic part of the classical Markov chain. This gives an internal structure for the particle as well as the entanglement entropy value. In such a digital physics substrate, the particle walks in such a way as to maximize its entanglement entropy based upon the inherent memory of its walking path (a measurement of the system on n implies the state at previous m), which generates an entropic force.

Interestingly, we can consider von Neumann Entropy (14) in the context of quantum information. The probabilities are the Markov chain connecting the steps. Accordingly, the particles construct "letters" of a spatiotemporal code as expression of the allowed restricted but non-determined walking paths, where the entropy measures the information needed. From this, we see how the entropic force emerges.

3 Results

3.1 Entropy of Black Hole

The framework above is a discretized quantum field expressing phases as algebraically allowed patterns upon fixed discrete geometry. Specifically a moduli space algebraic stack. Gravitational entropic force suggests a unified picture of gravity and matter via a quantum gravity approach. Consider now a regime from pure quantum gravity, like the black hole quantum horizon, so that there are no quantum fields but only quantum geometry. We can use our DQW to simulate this regime.

Again, DWQs encode local entanglement entropy (14) in the sense that the chains or network of step paths are always dynamically under construction as a "voxelated" animation. This resembles the isolated quantum horizons formulation of LQG [28, 29, 30, 31, 32], which gives explains the origin of black hole entropy in LQG. In this scenario, the Horizon area emerges from the stepwise animated actions at the Planck scale that are simulated by DQWs. We argue that DQWs are the Planck scale substrate forming the emergent black hole quantum horizon, where particle masses composite to black hole mass in the full aggregate of quantum walks on the spin network of the event horizon.

In the isolated quantum horizons formulation, entropy is generally calculated by considering the eigenvalues of the area operator $A(j)$ and introducing an area interval $\delta a = [A(j) - \delta, A(j) + \delta]$ of the order of the Planck length with a relation to the classical area a of the horizon. $A(j)$ is given by

$$A(j) = 8\pi\gamma l_p^2 \sum_l \sqrt{j_l(j_l + 1)}, \tag{1}$$

where γ is the Barbero-Immirzi parameter and l_p the Planck length. The entropy, in a dimensional form, is

$$S_{BH} = lnN(A), \tag{2}$$

with $N(A)$ the number of micro-states of quantum geometry on the horizon (area interval a) implemented, combinatorially, by considering states with link sequences that implement the two conditions

$$8\pi\gamma l_p^2 \sum_{l=1}^{N_a} \sqrt{j_l(j_l+1)} \leq a, \tag{3}$$

related to the area, where N_a is the number of admissible j that puncture the horizon area and

$$\sum_{l=1}^{N_a} m_l = 0 \tag{4}$$

related to the flux with m_l, the magnetic quantum number satisfying the condition $-j_l \leq m_l \leq j_l$.

The detailed calculation [29, 32] shows that the dominant contribution to entropy comes from states in which there is a very large number of punctures. Thus, it is productive to interpret this entropy as quantum informational entropy (14). Let us investigate how the horizon area and related entropy emerges from maximal entanglement entropy of DQWs. Condition (3) is associated with these DQWs. From (14 and 15), considering edge coloring, maximal entanglement entropy occurs for states on nodes of large valence N_{max} and sequence with $j_l = \frac{l}{2}$ ($l = 1, 2, ..., N_{max}$). Accordingly, for the entropy calculation, it is admissible that j, which punctures the horizon area, respects these sequences and that associated nodes have large valence rank. Therefore, we can rewrite condition (3) as

$$\sum_{i=1}^{N_{a_i}} a_i = a_c, \tag{5}$$

where N_{a_i} is the number of admissible nodes with

$$a_c = \frac{a}{4\pi\gamma l_p^2}. \tag{6}$$

and each a_i is calculated from

$$a_i = \sum_{l=1}^{N_{max}} \sqrt{l(l+2)}, \tag{7}$$

and considering the dominant contributions given by the over-estimate each a_i and counting $N(a_c)$ that will give $N(A)$, each a_i is an integer strictly greater than 1

$$\sqrt{l(l+2)} = \sqrt{(l+1)^2 - 1} \approx l+1, \tag{8}$$

which means that the combinatorial problem we need to solve is to find $N(a_c)$ such that (5) holds. This was discussed in a similar problem in [28]. It is straightforward[3] to see that

$$logN(A) = \frac{log(\phi)}{\pi\gamma} \frac{a}{4l_p^2},$$ (9)

where $\phi = \frac{1+\sqrt{5}}{2}$ is the golden ratio[4]

With this established, we can now conjecture that this entropy, relating to states that give maximal entanglement entropy, is simply the entanglement entropy of the DQWs. Accordingly, holographic quantum horizons can be simulated by the DQWs with maximal internal entanglement entropy. The walker moves are from node m to node n, each each with large valence number so that equation (16) works. If we consider a walker mass spanning I with Planck mass m_p, we can choose the proportionality constant so that

$$\triangle S_E = |S_{E_n} - S_{E_m}| = I,$$ (10)

so $M = Im_p$ gives the amount of information on the horizon, and $\triangle S_E$ is given in *bits*. We can make explicit the log_2 on (9) so that it is given in *bits* too and propose that $log_2 N(A) = \triangle S_E$ and that this measure of information gives the emergent Bekenstein-Hawking entropy. DQWs encode or express black hole information, so to infer its entropy, one needs to make contact with some frame of DQWs. The particles that simulate black holes will be more probable on a node of maximal entanglement entropy encapsulating $logN(A)$. Note that $logN(A)$ is not the usual statistical entropy because we have not taken into account all of the micro-states or the flux condition. From (2) the black hole entropy, changing the logarithm, is

$$S_{BH} = log_2 N(A) = \frac{log_2(\phi)}{\pi\gamma} \frac{a}{4l_p^2}.$$ (11)

Therefore, Bekenstein-Hawking entropy is recovered by setting the Barbero-Immirzi parameter

$$\gamma = \frac{log_2(\phi)}{\pi},$$ (12)

showing that the states of maximal entanglement entropy are dominant in black hole entropy. So we can think of Bekenstein-Hawking entropy as emergent from the local entanglement entropy above. A horizon area a spanning $4I$ Planck areas has I *bits* like (10).

---

[3]Because the cardinal $N(a_c)$ of the set of ordered tuples of integers strictly greater than 1 summing to a_c is the a_c^{th} Fibonacci number $F(a_c)$.

[4]The golden ratio [33] is an irrational number (1.61803398875...) as the solution to the equation $\phi^2 = \phi + 1$.

3.2 A model of walker position topologically encoded on a spin network

The Clebsh-Gordan condition at each node is realized by covering the graph with loops. From (3) and (14), we can compute the local entropy from a vertex as

$$S_{E_n} = log\sigma - \frac{1}{\sigma}\sum_{m}^{N} j_{mn}(j_{mn}+1)log\left(j_{mn}(j_{mn}+1)\right), \qquad (13)$$

where $\sigma = \sum_{m}^{N} j_{mn}(j_{mn}+1)$ of neighbor links. For example, at a node $\{2,3,3\}$, $j = \{1,\frac{3}{2},\frac{3}{2}\}$ gives $\sigma = \frac{19}{2}$ and $S_{E_n} = 1.06187$. At a node $\{2,2,2\}$, $j = \{1,1,1\}$ gives $\sigma = 6$, $S_{E_n} = 1.09861$, which is the maximum possible local entropy. Note that (13) is the local entropy formula given in [20] as 14 and is in accordance with well known LQG formulas for quantized length and area. See figure (8).

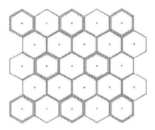

Figure 8: Loops.

The local entropy at each node is color coded. From equation (16), a massless particle moves on the same color and a massive particle moves along constant absolute color differences.

In the figure (9) :
(blue)$\{l_1, l_2, l_3\} = \{0,1,1\}$ or $\{0,2,2\}$ (side effect)
(white) ... $\{1,2,3\}$
(yellow) ... $\{1,1,2\}$
(orange) ... $\{2,3,3\}$
(red) ... $\{2,2,2\}$.

Dynamics:
Photons can orbit on orange hexagons. No possible particles with constant mass. Possible travel of massive particle, with $mass = |S_{(red)} - S_{(orange)}|$, $(= 0.037)$ interacting with photons at some orange vertex.

The walker position or the presence of a particle at one node is encoded by a triangle. Its move is a couple of 3-1 and 1-3 Pachner moves on neighbor positions, piloted by the walk probability. See figure (10).

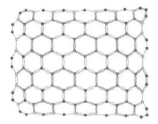

Figure 9: Entropy is color coded.

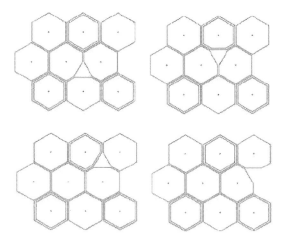

Figure 10: Particle and Pachner moves.

4 Ontological discussion

4.1 Minimal volume and charge

The generic implication of quantum gravity, the existence of a minimal volume (v_o) near the Planck scale indicates a new fundamental limit in nature. This fundamental limit that arises during the quantization of gravitational fields is similar to the fundamental charge q_e that arises in the quantization of electromagnetic fields.[5] For all these fundamental limits, there is one fundamental dimensionless constant, which relates them. It is often said that the dimensional constants are more fundamental than the dimensionless constants [34]. This is especially true when they are fundamental building blocks of a theory, such as occurs when there are general limits found in nature such the speed of light c in special relativity and the minimal action $\hbar$ in quantum mechanics. The dimensionless constants depend on the model, such as the standard model of particle physics, and they are generally determined ex-

---

[5]The same discussion can be made for fundamental charges associated with electroweak and the strong force that is in agreement with the understanding of a fundamental running coupling constant using renormalization group equations.

perimentally, where there is a meaningful experimental margin of error. We argue that there are dimensionless constants of nature that have a special and fundamental role as ratios of the fundamental limits of nature.

In quantum electrodynamics (QED), the fundamental ratio is the fine-structure constant α, wherein the relation is obtained from Dirac's quantization of magnetic charge [35]

$$\alpha = \frac{e}{g}n, \tag{1}$$

and where n is some integer, g is the magnetic charge rewritten in Coulombs and e the elemental electric charge. The magnetic charge is related to the non-trivial topology of charge space. Or, considering a discrete substrate, for each lattice we use to describe the electric field, there is a dual lattice corresponding to magnetic flux. So, for each site, we have the two fundamental charges living in respective dual spaces to one another. The charges of QED are quantized such that the ratio between any two is α, and the coupling constant is related to the strength of the force. We will not elaborate further here. However, we note that the monopole magnet provides a new understanding of the relation between topology, geometry and physics that is captured in the simplest form by equation (1).

For more insight, we can use a similar formula without flux by fixing to the Planck scale,

$$\alpha = e^2 \frac{k_e}{\hbar c} = \left(\frac{e}{q_p}\right)^2 \tag{2}$$

where k_e is Coulomb's constant or the electric force constant, a constant from the QED interaction, $\hbar$ is the reduced Planck's constant, c is the speed of light in vacuum and q_p is the Planck charge. In terms of α^{-1} ,

$$\alpha^{-1} = \frac{1}{e^2}\frac{\hbar c}{k_e} = \left(\frac{q_p}{e}\right)^2 . \tag{3}$$

The approximate value for this constant is based on experiments and QED perturbation theory calculations. The CODATA [36] recommended value is

$$\alpha = 0.0072973525664 \tag{4}$$

and

$$\alpha^{-1} = 137.035999139 \tag{5}$$

and is associated with the QED scale. It it important to take note of the fact that the CODATA value of α^{-1} in equation (5) is based on an average of disagreeing results given by five different experimental techniques for the measurement of Planck's constant. To explain, α is a ratio relative to the elementary charge. CODATA does not independently define the quantity for α. Instead, a value is derived from the relation (2). Accordingly, the accuracy of measurement of Planck's constant is proportional to our knowledge of the fine structure constant. The CODATA value for Planck's constant is a weighted average of the following five experimental techniques that agree only at the 4th place after the decimal $(6.6260 \times 10^{-34} Js)$. See figure 11 [36].

Method	Value of h (10^{-34} J·s)
Watt balance	6.626 068 89(23)
X-ray crystal density	6.626 0745(19)
Josephson constant	6.626 0678(27)
Magnetic resonance	6.626 0724(57)
Faraday constant	6.626 0657(88)
CODATA 2010 recommended value	**6.626 069 57(29)**

Figure 11: Planck's constant. CODATA 2010.

Experimentally, we can therefore only be confident in the value of h to the 4th decimal place, which limits our knowledge of what α is to the 4th decimal place. If one wishes to assume a probable value greater than the 4th decimal place, it would be logical to use the most advanced and precise experimental technique to date, as opposed to the CODATA approach that uses older lower resolution results to down-grade the precision of the more precise newer results. In the case of the gravitational constant, the precision is only to the 2nd decimal place. The most advanced high energy technique is based on atomic scale experiments not available when the earlier techniques averaged into the current CODATA value were developed. Specifically, the most technologically advanced calculation for measuring the gravitational constant was reported in Nature in 2014 [37]. It is a result derived by probing the atomic scale. Herein, we will refer to this as "highest resolution value of G", denoted as G_{HR}, $G_{HR} = 6.671 \times 10^{-11} m^3 kg^{-1} s^{-2}$.

In an attempt to discover what the actual minimum volume v_o is that nature uses, there have been many attempts to quantify it from string theory [38], loop quantum gravity, (LQG) [1] and others [16]. However, none of these unification theories have made successful predictions. So until a predictive quantum gravity theory is discovered, one cannot make solid theoretical assumptions using the CODATA value beyond a few decimal places for any fundamental constant such as α. When precision is desired, one would want to use G_{HR} combined with c to generate $\hbar$ and, from there, to generate the rest of the constants [39].

There is consensus that the quantum gravity scale is at or near the Planck volume. However, the limit of our knowledge of the value of $\hbar$ directly limits our knowledge to the current limit, which is G_{HR}. Fortunately, actual results permit us to understand that the gravitational constant, G, ties the three following fundamental limits of nature together to build a dimensionless fundamental constant that we call β, which relates to α. The three dimensional constants associated with these limits are the maximal local physical velocity, the speed of light c, the minimal action (or a minimal amount of information) $\hbar$ and this newly discovered minimal volume v_o. It is the fundamental dimensionless ratio defining the relationship between these fundamental dimensional limits. We write this as

$$\beta = \frac{1}{v_o^2} \left(\frac{\hbar G}{c^3} \right)^3 - \left(\frac{v_p}{v_o} \right)^2 \tag{6}$$

or in the inverse form

$$\beta^{-1} = v_o^2 \left(\frac{c^3}{\hbar G} \right) = \left(\frac{v_o}{v_p} \right)^2, \tag{7}$$

where we use the Planck volume, $v_p = l_p^3$, with l_p the Planck length that defines the Planck scale and Planck units as

$$l_p = \sqrt{\frac{\hbar G}{c^3}}. \tag{8}$$

We will discuss in the next sections how these constants β and α are deeply related to the golden ratio.

For more insight, let us focus on the quantization of the gravitational field using results from LQG [1]. Quantization is performed using a generalization of the lattice of QED — a triangulation of spacetime and its dual two-complex. Here we a split of the gravitational "charge" into two components; one related to flux[6] (gravitational analog of magnetic field) with area eigenvalues $l_p^2 \sqrt{j(j+1)}$ where j are non-negative half integers. And the other is the gravitational analog of the electric field in the dual space that is the geometry of quantized spacetime, which we will call l_l

$$l_l^2 \propto \gamma l_p^2 \sqrt{j(j+1)}, \tag{9}$$

where γ is the Barbero-Immirzi parameter. The proportionality constant in this reference is 8π. From this quantization condition, we derive the eigen-

---

[6]A less discussed quantity in the literature which is used for dynamics in quantum gravity formalisms are the volume flow rate, well known in hydrodynamics and magneto hydrodynamics, related to constant $Q_\Phi = \frac{4\pi\hbar G}{c^2}$ [40] given in the SI units by $m^3 s^{-1}$. Following the discussion above about the most accurate value of Planck's constant based on the most precise experimental technology, we reiterate that the CODATA values agree only at the 4th place after the decimal and the value for G agrees only to the 2nd place after the decimal. Using the result of the most recent and high energy atomic scale experimental techniques, G_{HR}, together with the value of h only to the 4th place after the decimal gives $Q_\Phi = 0.6180382 \times 10^{-59} m^3 s^{-1}$. This results in a value going out to the 5th decimal place with $1/\phi$.

value of the volume to be

$$v_0^2 \propto \gamma^3 \left(\frac{\hbar G}{c^3}\right)^3,$$

(10)

which gives

$$\beta \propto \frac{1}{\gamma^3}.$$

(11)

The proportionality constant in this reference is $6\sqrt{3}/(8\pi)^3$. The theory has some ambiguities and does not permit us to calculate this constant or γ. There are no experimental solutions possible yet. However, black hole calculations based on the limit of general relativity and quantum mechanics provide an approximate value for γ between 0 and 1. See equation (12) and reference [32].

This indicates a relationship between β and the quantum gravity coupling constant, γ, associated with the gravitational constant. That is, β leads to the precise quantum gravity coupling constant—also a number between 0 and 1.

Correspondence between the two scales, QED and quantum gravity, is expected in a correct unification theory. Our interpretation is that it is related to the quantization of charge and is manifest in the ratios of equations (1) and (9). The result suggests an underlying principle governing this fundamental ratio. We suggest this principle is related to symmetry and graph network theory efficiency, where nature exists as a maximally efficient graph theoretic topological quantum code due to its requirement of classic efficiency as exemplified in the principle of least action and other conserved symmetries ubiquitous in nature, such as the gauge symmetry unification of particles and forces in the standard model of particle physics. Put differently, presume the digital physics view is true, where nature is information theoretic. That is, nature is based on symbolic language and not merely described by symbolic language. From the probability plot in 3-space of the quantum wave form, to gravity theory to classic physics, this supposed language is simulating or expressing itself geometrically in the form of physics. Accordingly, a geometric code would be a good conjecture for what form of symbols would be used in the mathematical language of our geometric universe. And because nature appears to be concerned with efficiency, a maximally efficient geometric code would be logical. Graph drawing codes, such as phason algebraic codes in quasicrystals, may be optimally efficient in the universe of codes capable of expressing a geometric physical universe.

The efficiency of a code, such as a geometric spatiotemporal code, can be ranked by the ratio of binary actions needed to express a given physical system. A geometric based quantum gravity plus particle physics code would express spatiotemporal information using the minimum quantity of irreducible geometric symbols. Trivially, the minimum number of symbols in any code is two. Fundamental quasicrystals such as the Elser-Sloane quasicrystal [41] with H_4 symmetry, derived by projective transformation of a minimal slice of the E_8 lattice, or the quasicrystalline spin network (QSN) [42, 43] with H_3

symmetry are constructed as geometric languages composited from geometric spatiotemporal symbols as points separated by two distances related as 1 and $1/\phi$. Such codes exist physically in nature as quasicrystalline phases of matter, where generally a network of Fibonacci chains [15] in 3-space manifests as energy wells separated by 1 and $1/\phi$ instead of abstract points in the mathematically ideal abstract analogue. At any given moment, some energy wells are occupied by an atom and some are vacant. Over the time domain, the atoms are known to tunnel at very low energy according to a phason code that is inherently non-local.

The binary efficiency of a network can be ranked by the connectivity rank of its nodes—specifically the average valence value of its vertices. Maximally connected graph drawings can only be achieved via the use of the golden ratio, due to it being the only ratio possessing the fractally self-similar quality where $\dfrac{A}{B} = \dfrac{A+B}{A}$, where $A = 1$ and $B = \dfrac{1}{\phi}$.

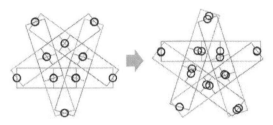

Figure 12: To optimally network 1D two-letter codes in 2, 3 and 4 dimensions, we must use the Fibonacci chain with the two letters as $1/\phi$ or the connectivity breaks down. An infinitesimal deviation from the golden ratio spacing renders the object (1) not a quasicrystal because arbitrary closeness of nodes will exist and (2) a non-code because there will be an infinite number of 1D symbols as lengths and (3) not a quantum topological network because non-local influence where an action on one point influences other non-local point on/off states breaks down. We know that the reason for this is because, in the universe of all ratios closed under multiplication and division, only ϕ possesses fractal self-similarity. The above diagram on the left shows a network of five short Fibonacci chains, where nodes overlap perfectly due to the fractal self-similarity of ϕ. To left, we see how a small deviation from the golden ratio spacing destroys the connectivity rank by lowering the average near neighbor valence magnitude and introducing arbitrary closeness between nodes.

Our conjecture is that this fundamental dimensionless constant β discussed above is $\sqrt{\beta} = 1/\phi^3$, which makes the ratio between the Planck length and the minimum length l_o, related to v_o, ϕ and allows for networks where nearest neighbor lengths are restricted only to 1 and ϕ, starting at the Planck scale. Due to its power of network code efficiency, this fundamental ratio would be the building block of the standard model of particle physics and general relativity in a geometric graph-drawing code theoretic unification formalism. Note that in a gauge unification picture, Lie groups, Lie algebras and their associated lattices, such as the E_8 lattice, play a central role [44, 45]. Grand unified theories are elegant and successful. They embed the Standard Model Lie group inside larger groups, where in 6D the largest group possible is $E6$ [38, 46]. Going to higher dimensions, $E6$ is a subspace of $E8$.[7] $E8$ is a powerful tool unifying gravity and the standard model via the

---

[7] $SU(3) \times E6$ the maximal sub algebra of $E8$.

most popular and foundational form of string theory, heterotic string theory [38, 47, 48] or $E8$ Grand Unification by itself [49, 50, 51, 52, 53]. Remarkably, $E8$ symmetry has been found in condensed matter physical experiments [54]. Furthermore, one can recover $E8$ gauge symmetry physics from an icosian construction of the E_8 lattice [55].[8]

Summarizing this section, if nature is a geometric code, such as a quasicrystal phason code derived from the higher dimensional E_8 lattice, this might explain charge quantization in nature and shed light on a universally generalized optimization principle. Let us discuss this in detail next.

4.2 The principle behind fundamental ratios

To discover the correct quantum gravity and particle physics unification theory, new insights and fundamental principles must be realized [56, 57]. The code theoretic axiom may lead us in the right direction [2]. The existence of a minimal volume implies violation or generalization of principles such as Lorentz transformations, the uncertainty principle and the equivalence principle [58, 59, 60, 61]. We argue for a new principle associated with this minimal volume, a dimensionless constant.

It is trivially true that ratios are more fundamental than numerical values based on arbitrary metrics. If we consider a model of quantum geometry, as developed in various quantum gravity theories, these ratios will be geometric as will any building blocks of a realistic model. If this is the case, what is the principle governing the organization of these building blocks at the Planck scale? Or more generally, what is the principle behind charge and, equivalently, length and volume quantization that can be deduced by looking at the ratios of fundamental limits:

$$v_p = \sqrt{\beta} v_o \tag{12}$$

and,

$$e = \sqrt{\alpha} q_p \tag{13}$$

In Planck units:

$$v_p = 1$$
$$v_o = 1/\sqrt{\beta}. \tag{14}$$

For the elementary electric charge

$$q_p = 1$$
$$e = \sqrt{\alpha}. \tag{15}$$

So this building block ratio for charge and volume quantization is a specific fundamental ratio based on the efficiency and symmetry principle we are looking for.

---

[8]Note that geometrically, E_6 is a sub-lattice embedded in E_8.

It is generally agreed that fundamental physics relies on a fundamental optimization principle, like we referenced in [62]: "The whole of theoretical physics (classical and quantum) relies on a fundamental optimization principle from which the basic equations of physics can be constructed under the form of Euler–Lagrange equations, relating fundamental quantities, such as energy, momentum and angular momentum. This principle of least action becomes a geodesic principle in the framework of relativity theories (i.e., the action is identified with the proper time). The geodesic principle states that the free trajectories are the geodesics of spacetime. It plays a very important role in a geometric relativity theory, since it means that the fundamental equation of dynamics is completely determined by the geometry of spacetime and therefore does not need to be set down by an independent equation. Moreover, in such a framework, the action can be identified (modulo a constant) with the fundamental metric invariant, which is nothing but the proper time itself. The action principle becomes nothing else than the geodesic principle. As a consequence, its meaning becomes very clear and simple: the physical trajectories are those, which minimize the proper time itself".

So the principle of least action is a principle related to efficiency and optimization of resources and has a defined meaning at a classical level; the minimal path. That is, the action is taken as a stationary value. Here we have a clear first-principles understanding of "efficiency". A similar optimization principle can be found in computational systems in the principle of least computational action [63]. Putting together the least action and relativity principles, we are lead to a principle of maximally efficient ratio. However, we are interested in discrete spacetime implied by quantum gravity. Accordingly, at the quantum level, the classical notion of a sharp trajectory followed by a physical system is rejected, leaving an opportunity. Specifically, the ratios defining the discrete geometry at the level of quantum gravity. To make this principle of maximally efficient ratio precise, we will define the discrete substrate of quantum gravity. Before doing so, let us articulate the principle in the context of discrete quantum geometric evolution:

↪ The evolution of a system is such that it uses the most efficient ratio relating its indivisible building blocks.

In section 2.1, we considered a model for a particle interacting with a discrete substrate of the gravitational field, a fixed spin network from LQG. The interaction is given by equation (1). The interaction term takes this form because the relevant Hilbert space depends on the wave functions at the vertices. From this, we derive the dynamics by implementing a unitary evolution with a quantum random walk. Here we will go further to propose that the transition from a node v_m to a node v_n is guided by one action $S(v_n, v_m)$. Equation (1), in fact, implies a Laplacian operator for the graph because we have a discrete structure that derives energy values. Classically, this action is a function that integrates the two steps leading from v_m to v_n, i.e., between the two discrete times. In order to account for such a discrete substrate, the action must be considered probabilistically. From random walks on a graph theoretic spin network, we derive the probability $P(v_n)$ for a particle ending up on an arbitrary vertex v_n along with the transition probabilities $P(v_n, v_m)$ for transitions from v_m to v_n. That is, the probabilities are used to construct the operators for unitary evolution. Considering the action probabilistically, the average action is calculated, which can be achieved locally by $\sum_{v_m} P(v_n, v_m) S(v_n, v_m)$ and globally in the spin network Γ by considering that random walks on arbitrary vertices generate a global average action S_Γ

$$S_\Gamma - \sum_{v_n} \sum_{v_m} P(v_n) P(v_n, v_m) S(v_n, v_m). \tag{16}$$

This is the total average action. To find the most efficient ratio between spacetime and particle building blocks, we must look for the probabilities that minimize the action. This is intuitive because probabilities are ratios. And the sum above implies the existence of a set of probabilistic matrices. It is interesting to note that both the Pauli and Dirac matrix formalisms can be reduced to binary matrices. The highest probability non-trivial eigenvalues for any $n \times n$ binary matrix are ϕ and $-1/\phi$. See Appendix (7). The precise definition of "non-trivial" depends on the dimension. At small dimensions (less than 10), the domination of ϕ is very clear. After dimension 50, additional "trivial" values can be the most probable with ϕ still holding a relatively high rank. Naturally, when the dimension goes to infinite, only 0, the singular trivial value, has a significant probability, all other eigenvalues going to a zero measure.

The efficiency of discrete quantum geometry can be exploited by looking at which of those matrices describe realistic evolution and minimize the average total action. In other words, find the matrix generating the minimal action using approaches which can be implemented by a partition function in the language of statistical mechanics or a quantum mechanics path integral formalism. At this level, it is the same object. So this is an efficiency and optimization action principle that can work for quantum gravity and for discrete versions of other force quantization models. Further investigation is planned to make clear how to derive, from the action principle, (16) the building blocks for the charge and volume quantization equations (14) and (15). However, for now, the above evidence implies that these foundational matrices are deeply related to the golden ratio. See Appendix (7). The code theoretic physics view, which implies an information theoretic universe, indicates that the choice of fundamental ratios and building blocks must be guided by the notion of efficiency or computational least action [63]. This leads naturally to the golden ratio with its powerful properties of being deeply associated with icosians [64], gauge symmetry physics, closure under multiplication and division and the crucial fractal self-similarity quality where $\dfrac{A}{B} = \dfrac{A+B}{A}$, allowing for maximally connected quantum-topological non-deterministic graph codes. Interestingly, because symbolism traffics in the quantity called information or meaning and, considering the principle of efficient language [2], we see that the most principle may be one that takes into account the rank of geometric meaning, which can be considered as complexity of order. For example, in the entropic phase of some binary alloy, we can have a high entropy phase with low complexity of order when order is low. And, at the low entropic phase, we have high oder at the crystal phase but low complexity rank because the network of objects is a homogeneous arrangement with all near neighbor distances being equal. Code theoretically, this is an ordering of just one spatiotemporal symbol, which defines it as a non-code. The maximum rank of complexity order is at the non-zero limit of spatiotemporal symbols, which is 2. The only way to achieve this in a network is by compositing a network of 1D two-letter codes known as Fibonacci chains make of spacings that are 1 and the golden ratio.

$\hookrightarrow$ If reality is code theoretic, its purpose is to efficiently express meaning, such as physical information, with its ultimate conserved quantity—quantized actions of the evolving substrate. Specifically, syntactically free binary choices in the self-emergent code theoretic network. Efficiency is achieved by (1) operating as a neural network code that generates maximal meaning from binary actions and (2) strategically placing those choices for maximal meaning (generally, physical meaning) according to syntactically free code choices.

4.3   Golden ratio

From the information theoretic and code theoretic views, particles, motions, interactions and spacetime geometry are language based. This new paradigm implies a non-arbitrary code operating at the Planck scale. We suggest the problem of quantum gravity is the search for the discovery of this non-arbitrary and efficient code of nature. With respect to physical codes, much progress has been made in understanding quasicrystalline codes. The most famous is the DNA code [65, 66], a golden ratio based quasicrystal biomolecule. Quasicrystal codes in materials science have also been studied [15, 42, 43, 67, 68, 69, 64, 70].

In 1944, Schrödinger predicted that the molecule encoding life would be a "quasiperiodic crystal" before the term quasicrystal was coined in the 1970s [71]. Amazingly, this was well before the precise molecular identification by Crick and Watson [66]. Each of the two DNA strands is a 5-periodic helix exactly expressed with the golden ratio. The two strands offset along the shared helical axis by two sequential Fibonacci numbers closely approximating the golden ratio. DNA has rotational symmetry but not translational symmetry—one of the defining characteristics of quasicrystals. DNA is a 3D network of deep and narrow double well potentials allowed by golden ratio based atomic organization. This aperiodic discrete energy landscape is part of what defines DNA code, where efficiency is achieved via the maximum spatiotemporal restriction of atoms without being fully restricted, as opposed to a crystal wherein all energy wells of the same size are occupied in each atom is locked into an EM trap. That is, a quasicrystal is a network of double well potentials with some wells occupied and others vacant, wherein spatiotemporal freedom of the atoms approaches the non-zero limit. For example, unlike a crystal, the assembly rules for a quasicrystal allow self-organization or construction choices within the rules that are not forced, i.e., non-deterministic. And dynamically, the orderly pattern of vacant energy wells allows particle dynamics not possible in crystalline or amorphous phases of matter. Due to the fractal self-similar quality where $\frac{A}{B} = \frac{A+B}{A}$, the non-zero limit of spatiotemporal freedom can only be achieved via golden ratio spacing ratios between energy wells.

A quasicrystal is a structure that is ordered but not periodic. It has long-range quasiperiodic translational order and long-range orientational order. It has a finite number prototiles or "letters". And it has a discrete diffraction pattern indicating order but not periodicity. Mathematically, there are three common ways of generating a quasicrystal: the cut-and-project method (projection of an irrational slice of a higher dimensional crystal) [15], the dual grid method [15], and the Fibonacci grid method [42, 43]. Finite quasicrystals can be constructed by matching rules. Quasicrystals were discovered in nature via synthesis only in 1984 [72].

A hallmark and general characteristic of the 300 or so quasicrystals physically discovered is the golden ratio. Most of these quasicrystals are projections or subsets of the Elser-Sloane quasicrystal [65], which is a cut-and-project of the E_8 lattice using the angle $\arctan(1/\phi^3) \approx 13.28$ and or the angle $\pi/4 - \arctan(1/\phi^3) \approx 22.24$. The simplest quasicrystal possible is the two length Fibonacci chain, as 1 and $1/\phi$.

The Penrose tiling, a 2D quasicrystal, is a network of 1D quasicrystals. 3D quasicrystals in nature, such as a 3D Penrose tiling (Ammann tiling) are networks of 2D quasicrystals, which are each networks of 1D quasicrystals—generally Fibonacci chains. So the irreducible building block of all quasicrystals are 1D quasicrystals. The "letters" of these 1D spatiotemporal codes are lengths between vacant or occupied energy wells. And a 1D quasicrystal can have any finite number of letters. However, the minimum is two. The Fibonacci chain is the quintessential 1D quasicrystal. It possesses two lengths related as the golden ratio. There are an infinite number of 2-letter 1D quasicrystals. However, maximally code efficient quasicrystals in higher dimensions that utilize only two letters must be constructed from Fibonacci chain 1D quasicrystals with letters being 1 and the inverse of the golden ratio as uniquely generated by a cut-and-project of the Z_2 lattice, in the direction of $\arctan(1/\phi^3) \approx 13.28$ degrees from the diagonal direction of the unit cubic cells.

Returning to the principle of efficient ratios and quantization conditions (1) and (9), we note the complementary role of the dual space in the case of discrete structures like DNA and quasicrystals in general. In (1) and (9), this relationship is obvious.[9] This dual space can be derived from the first object by generating its diffraction pattern. The details of the duality and quantization condition depend on the specific physics considered. The diffraction

---

[9]Let $\mathcal{L}$ be a lattice. All lattices $\mathcal{L}$ have an associated dual lattice $\mathcal{L}^*$, the set of vectors $\vec{y}$ whose scalar products with the vectors $\mathcal{L}$, $\vec{x}$ are integers, $\vec{x}.\vec{y} = n$.

pattern is quantized in relation to the distance between physical objects in physical space. The interesting case is when distances are irrational numbers such that the ratios allow aperiodic order evidenced by the discrete Bragg peaks in the dual space as the diffraction pattern. Again, when the irrational number is the golden ratio, values are closed under multiplication and division and the ratio itself is fractally self-similar, which allows the maximally connected or densest network of 2-letter quasicrystals to be organized in dimensions 2D to 4D—the convergence of networks of 1D Fibonacci chain codes into higher dimensional quasicrystal codes.

$\hookrightarrow$ Let us bring together some of the above ideas to support the conjucture that the coupling constant at the Planck scale is $1/\phi^3$ —the quantization of charge and the proposed discretization of spacetime. Simplex integers, as shape-numbers, are the fundamental building blocks of the QSN quasicrystal (a 3D network of Fibonacci chains) [14]. The principle of efficiency of ratios or the more general principle of efficient language states that fundamental ratios allow maximally efficient geometric codes, specifically quasicrystals. We presented the prominence of the golden ratio in two important natural codes—the DNA quasicrystal and metallic quasicrystals.

Next, we shall present some results in fundamental physics that may be clues that this direction is correct.

4.4   Golden ratio in physics

In this section, we review results that come from different fields of physics. All indicate the preeminence of the golden ratio. These results suggest that, with theoretical and experimental advances, the role of the golden ratio will become more clear.

Atomic physics—the hydrogen atom

The hydrogen atom is important because it is the simplest and most abundant element, having a single electron and nucleon. It serves as the foundation for all atomic theory. With the hydrogen atom, we have a case where the electron and atomic charge are identical, allowing us to isolate fundamental atomic structure ratios to test for the principle of efficient ratios. First, let us consider the hydrogen spectral series. When one electron goes from a higher to lower energy state, spectral emission occurs. The wavelengths of emitted/absorbed photons is given by the Rydberg formula in vacuum

$$\frac{1}{\lambda_{\text{vac}}} = R \left(\frac{1}{n_1^2} - \frac{1}{n_2^2} \right), \tag{17}$$

where R is the Rydberg constant

$$R = \frac{\alpha^2 m_e c}{4\pi\hbar} = \frac{\alpha^2}{2\lambda_e}, \tag{18}$$

and n_1 and n_2 are integers greater than or equal to 1 such that $n_1 < n_2$ corresponding to the principal quantum numbers of the orbitals occupied before and after the quantum leap. For example, with $n_1 = 1$, we have the Lyman series [73], which we can rewrite as

$$\lambda_{vac} = \frac{1}{R} \left(\frac{n_2^2}{n_2^2 - 1} \right) = \frac{2\lambda_e}{\alpha^2} \left(\frac{n_2^2}{n_2^2 - 1} \right), \tag{19}$$

where λ_e is the Compton wavelength of the electron. It gives for $n_2 = 2$. For example, $1.2156 \times 10^{-7} m$. If we use an approximation as $\alpha = \frac{1}{20\phi^4}$

$$\lambda_{vac} = 800\phi^8 \lambda_e \left(\frac{n_2^2}{n_2^2 - 1} \right) = 800(13 + 21\phi)\lambda_e \left(\frac{n_2^2}{n_2^2 - 1} \right), \tag{20}$$

we get $1.2158 \times 10^{-7} m$. As expected, the first excited state of the hydrogen atom is ϕ-based [74]. When we add corrections of the energy levels of the hydrogen atom due to relativity theory and spin-orbit coupling, the hydrogen fine structure shows up. The fine structure is α-based and ϕ-based.

Standard Model: Neutrino Mixing and the Cabibbo angle

The first evidence for physics beyond the standard model (SM) of particle physics comes from experimental neutrino oscillations. See [75] and references therein for a review. One case of experimental evidence is the detection of solar neutrinos. Only electron neutrinos are emitted by the sun. However, only about 30 percent of the number predicted by theories that explain how the sun works are actually measured. This disagreement was resolved by an improved understanding of neutrino physics. Specifically, electron neutrinos are converted into muons and tau neutrinos. This is in agreement with various experiments. Neutrino oscillation is a verification that neutrinos are massive, which is in disagreement with the SM, where neutrinos are expected to be massless. So the understanding from experiment is that there is a lepton mixing matrix U—the Maki-Nakagawa-Sakata (MNS) matrix U_{MNS}, which relates the basic SM neutrino states, ν_e, ν_μ, ν_τ, related with the electron, muon and tau to the neutrino mass states ν_1, ν_2, and ν_3 with masses m_1, m_2, and m_3.

$$\begin{pmatrix} \nu_e \\ \nu_\mu \\ \nu_\tau \end{pmatrix} = \begin{pmatrix} U_{e1} & U_{e2} & U_{e3} \\ U_{\mu 1} & U_{\mu 2} & U_{\mu 3} \\ U_{\tau 1} & U_{\tau 2} & U_{\tau 3} \end{pmatrix} \begin{pmatrix} \nu_1 \\ \nu_2 \\ \nu_3 \end{pmatrix} \tag{21}$$

This is similar to the usual CKM (Cabibbo–Kobayashi–Maskawa) quark mixing matrix. The lepton mixing matrix U_{MNS} has large uncertainties. Despite the lack of a deep knowledge of U_{MNS}, the aforementioned data leads to an approximation of the first order, which allows for a theoretical description. There are different patterns of the values and parameterizations for this matrix that are in agreement with experimental data.[10] We suspect that the most powerful mixing matrix models are golden ratio based, as shown in [76] and was further developed in [77, 78, 79, 80], the so called golden ratio prediction. But more important than the exact values of the numbers that constitute the matrix is the idea that they can be derived via first principles from the symmetry breaking relationships of fundamental symmetries at the unification scale relating to hyper-dimensional lattices and their associated Lie algebras. For example, one can use the rotational icosahedral group [78], the alternating group of five elements A_5 [79] or $A_5 \times Z_5 \times Z_3$ [80]. And one can recover gauge symmetry physics from an icosian construction of the E_8 lattice [55]. Of course, by definition, all these groups are golden ratio based.

In this context, we have the important concept of quark-lepton complementarity which proposes that one parameter of the U_{MNS} matrix, the solar angle (θ_{12}), is related with one parameter of the quark matrix mixing, the Cabibbo angle (θ_c), by

$$\theta_{12} + \theta_c = \frac{\pi}{4}. \tag{22}$$

We conjecture that the Cabibbo angle is the universal parameter which controls the entire structure of fermion masses and therefore appears in many places, such as mass ratios and mixing parameters. Recent data and theoretical proposals indicate the possibility of $\theta_c \approx 13.28$ [75, 77, 81, 82, 83]. This fits well with the golden ratio prediction [77]. The findings of these authors is not surprising because, as mentioned earlier, this angle appears in the cut-and-project of E_8 to 4D, breaking the crystal symmetry of E_8 to icosian associated H_4 symmetry. Similarly, this angle is required to bring the projection of E_8 to 3D into H_3 symmetry.

Both the 4D and 3D projections are golden ratio based quasicrystals. Similarly, the Weinberg angle (θ_W) of the electroweak interaction is an unknown value experimentally suggested to be slightly less than $\sim 30°$. In practice, what is measured is the quantity $sin^2(\theta_W)$. In a study of parity violation in Møller scattering a value of $sin^2(\theta_W) = 0.2397 \pm 0.0013$ was obtained at momentum transfer, $Q = 0.16 GeV/c$, establishing experimentally the "running" of the weak mixing angle. LHCb measured, in 7 and 8 TeV proton-proton collisions, an effective angle of $sin^2(\theta_W) = 0.23142$, though the value of Q for this measurement is determined by the partonic collision energy. The mean of these measured values deviates from $1/\phi^3$ at five 10,000ths. The measurement uncertainty of 0.2397 is 0.0013, the 1,000ths part, so our conjecture is well within the experimental margin of error.

To understand the origin of fermion masses and mixtures in the SM, one must turn to the family of symmetries that restrict the form of the mass generator matrices and models the hierarchies and mixtures through small symmetry breaking perturbations. Recent investigations, such as in the above references, indicate that icosahedral symmetry acts as an important scale unifier of the electromagnetic, weak and strong forces ($\sim 10^{25} eV$). If flavor has an underlying simplicity associated with unification symmetry, it

---

[10]Different constructions of this matrix can be found in [75].

would be productive to attempt recognition of it from neutrinos instead of from charged leptons and quarks, which have more complicated structures. This suite of experimental and theoretic evidences for massive neutrinos and our theoretical proposal of quasicrystal symmetry in unification scale physics supports the conjecture of a quasicrystalline spacetime and particle structure at the Planck scale.

We have found fundamental golden ratio values in the structure of E_8 itself. The root vector polytope of E_8 is the Gosset polytope.[11] One of the building blocks of E_8 is the 3-simplex, a regular tetrahedron. For example, the E_8 lattice and the Gosset polytope can be constructed, if one chooses, solely from tetrahedra. Two face-kissing tetrahedra in E_8 live in the same 4D subspace of R^8. And certain pairs share one vertex that is the centroid of the Gosset polytope while their other vertices are the vertices of the Gosset polytope. The following table 4.4 shows the list of the cosine values of the angles, most of which are golden ratio expressions. Notice that the angle $ArcCos\frac{1}{4}$ can be written as $\pi/3 - ArcCos - \frac{3\phi-1}{4}$. It is worth noting that every eight of these tetrahedra live in the same 3-space (grouped into two dual orientations) and can be extended to an A_3 lattice. Since, the Gosset polytope can be generated by composition of all the transformations of one tetrahedron through these angles, the E_8 lattice can be generated via composition of the rotational transformation of the A_3 lattices through these angles.

$\frac{1}{2}$	$\frac{7}{(4\phi-2)^2}$	$-\frac{3}{8\phi-4}$	$\frac{1}{4}$	$\frac{1}{2(2\phi-1)}$	$\frac{2\phi-1}{4}$

The Cartan matrix of the E_9 Lie affine algebra and the E_8 Lattice is:

$$
C_{E_9} = \begin{pmatrix}
2 & -1 & 0 & 0 & 0 & 0 & 0 & 0 & 0 \\
-1 & 2 & -1 & 0 & 0 & 0 & 0 & 0 & 0 \\
0 & -1 & 2 & -1 & 0 & 0 & 0 & 0 & 0 \\
0 & 0 & -1 & 2 & -1 & 0 & 0 & 0 & 0 \\
0 & 0 & 0 & -1 & 2 & -1 & 0 & 0 & 0 \\
0 & 0 & 0 & 0 & -1 & 2 & -1 & -1 & 0 \\
0 & 0 & 0 & 0 & 0 & -1 & 2 & 0 & 0 \\
0 & 0 & 0 & 0 & 0 & -1 & 0 & 2 & -1 \\
0 & 0 & 0 & 0 & 0 & 0 & 0 & -1 & 2
\end{pmatrix}
$$

Its 9 eigenvalues are $4, \phi+2, 3, \phi^2, 2, 2-\phi^{-1}, 1, \phi^{-2}, 0$ and form the diagonal matrix Λ_{E_9}. The last eigenvalue is 0 because the Lie algebra is affine. Five of the eigenvectors (columns of V_{E_9}) have integer coordinates and the four other have coordinates which are 0 or a power of ϕ. We note φ for the inverse of ϕ.

---

[11] The E8 root system contains 240 root vectors spanning R^8, each with the same length and forming the vertices of the Gosset polytope.

$$
V_{E_9} = \begin{pmatrix}
-1 & \phi & -1 & \varphi & -1 & -\varphi & 1 & -\phi & 1 \\
2 & -\phi^2 & 1 & -\varphi^2 & 0 & -\varphi^2 & 1 & -\phi^2 & 2 \\
-3 & \phi^2 & 0 & -\varphi^2 & 1 & \varphi^2 & 0 & -\phi^2 & 3 \\
4 & -\phi & -1 & \varphi & 0 & \varphi & -1 & -\phi & 4 \\
-5 & 0 & 1 & 0 & -1 & 0 & -1 & 0 & 5 \\
6 & \phi & 0 & -\varphi & 0 & -\varphi & 0 & \phi & 6 \\
-3 & -1 & 0 & 1 & 1 & -1 & 0 & 1 & 3 \\
-4 & -\phi & -1 & -\varphi & 0 & \varphi & 1 & \phi & 4 \\
2 & 1 & 1 & 1 & 0 & 1 & 1 & 1 & 2
\end{pmatrix}
\qquad
\Lambda_{E_9} = \begin{pmatrix}
4 & 0 & 0 & 0 & 0 & 0 & 0 & 0 & 0 \\
0 & \phi+2 & 0 & 0 & 0 & 0 & 0 & 0 & 0 \\
0 & 0 & 3 & 0 & 0 & 0 & 0 & 0 & 0 \\
0 & 0 & 0 & \phi^2 & 0 & 0 & 0 & 0 & 0 \\
0 & 0 & 0 & 0 & 2 & 0 & 0 & 0 & 0 \\
0 & 0 & 0 & 0 & 0 & 2-\varphi & 0 & 0 & 0 \\
0 & 0 & 0 & 0 & 0 & 0 & 1 & 0 & 0 \\
0 & 0 & 0 & 0 & 0 & 0 & 0 & \varphi^2 & 0 \\
0 & 0 & 0 & 0 & 0 & 0 & 0 & 0 & 0
\end{pmatrix}
\tag{23}
$$

$$
D_{E_9} = \frac{1}{120} \begin{pmatrix}
1 & 0 & 0 & 0 & 0 & 0 & 0 & 0 & 0 \\
0 & 12\varphi^2 & 0 & 0 & 0 & 0 & 0 & 0 & 0 \\
0 & 0 & 20 & 0 & 0 & 0 & 0 & 0 & 0 \\
0 & 0 & 0 & 12\phi^2 & 0 & 0 & 0 & 0 & 0 \\
0 & 0 & 0 & 0 & 30 & 0 & 0 & 0 & 0 \\
0 & 0 & 0 & 0 & 0 & 12\phi^2 & 0 & 0 & 0 \\
0 & 0 & 0 & 0 & 0 & 0 & 20 & 0 & 0 \\
0 & 0 & 0 & 0 & 0 & 0 & 0 & 12\varphi^2 & 0 \\
0 & 0 & 0 & 0 & 0 & 0 & 0 & 0 & 1
\end{pmatrix}
\tag{24}
$$

$$
V_{E_9}^{-1} = D_{E_9} V_{E_9}
\tag{25}
$$

$$
C_{E_9} = V_{E_9}\Lambda_{E_9}D_{E_9}V_{E_9} = V_{E_9}\Lambda_{E_9}V_{E_9}^{-1}
\tag{26}
$$

$$
\begin{pmatrix}
2 & -1 & 0 & 0 & 0 & 0 & 0 & 0 & 0 \\
-1 & 2 & -1 & 0 & 0 & 0 & 0 & 0 & 0 \\
0 & -1 & 2 & -1 & 0 & 0 & 0 & 0 & 0 \\
0 & 0 & -1 & 2 & -1 & 0 & 0 & 0 & 0 \\
0 & 0 & 0 & -1 & 2 & -1 & 0 & 0 & 0 \\
0 & 0 & 0 & 0 & -1 & 2 & -1 & -1 & 0 \\
0 & 0 & 0 & 0 & 0 & -1 & 2 & 0 & 0 \\
0 & 0 & 0 & 0 & 0 & -1 & 0 & 2 & -1 \\
0 & 0 & 0 & 0 & 0 & 0 & 0 & -1 & 2
\end{pmatrix} =
$$

$$
\begin{pmatrix}
-1 & \phi & -1 & \varphi & -1 & -\varphi & 1 & -\phi & 1 \\
2 & -\phi^2 & 1 & -\varphi^2 & 0 & -\varphi^2 & 1 & -\phi^2 & 2 \\
-3 & \phi^2 & 0 & -\varphi^2 & 1 & \varphi^2 & 0 & -\phi^2 & 3 \\
4 & -\phi & -1 & \varphi & 0 & \varphi & -1 & -\phi & 4 \\
-5 & 0 & 1 & 0 & -1 & 0 & -1 & 0 & 5 \\
6 & \phi & 0 & -\varphi & 0 & -\varphi & 0 & \phi & 6 \\
-3 & -1 & 0 & 1 & 1 & -1 & 0 & 1 & 3 \\
-4 & -\phi & -1 & -\varphi & 0 & \varphi & 1 & \phi & 4 \\
2 & 1 & 1 & 1 & 0 & 1 & 1 & 1 & 2
\end{pmatrix}
\left\lfloor \begin{matrix} 4 \\ \phi+2 \\ 3 \\ \phi^2 \\ 2 \\ 2-\varphi \\ 1 \\ \varphi^2 \\ 0 \end{matrix} \right\rceil
\begin{pmatrix}
-1 & \phi & -1 & \varphi & -1 & -\varphi & 1 & -\phi & 1 \\
2 & -\phi^2 & 1 & -\varphi^2 & 0 & -\varphi^2 & 1 & -\phi^2 & 2 \\
-3 & \phi^2 & 0 & -\varphi^2 & 1 & \varphi^2 & 0 & -\phi^2 & 3 \\
4 & -\phi & -1 & \varphi & 0 & \varphi & -1 & -\phi & 4 \\
-5 & 0 & 1 & 0 & -1 & 0 & -1 & 0 & 5 \\
6 & \phi & 0 & -\varphi & 0 & -\varphi & 0 & \phi & 6 \\
-3 & -1 & 0 & 1 & 1 & -1 & 0 & 1 & 3 \\
-4 & -\phi & -1 & -\varphi & 0 & \varphi & 1 & \phi & 4 \\
2 & 1 & 1 & 1 & 0 & 1 & 1 & 1 & 2
\end{pmatrix}^{-1}
\tag{27}
$$

Equation 25 gives a simple expression of the inverse of the eigenvector matrix. Equation 26 states the diagonalization of the Cartan matrix. Equation 27, where the formal notation $\lfloor v \rceil$ expresses the diagonal matrix form of the vector v, (here $\lfloor v \rceil = \Lambda_{E_9}$), is unexpectedly a new formula not previously reported in the literature.

Our intent was to look for a result in E_8 similar to that found by Sirag in the relationship between E_7 and $C_{E_7^+}$ [84]. We analyzed the spectrum of the C_{E_9} Cartan matrix to obtain our result— a systematical study of the spectrum of all Cartan matrices though the Dynkin diagram of all simple Lie algebras and their affine extensions. We found that the golden ratio appears in the eigenvalues, outside of the exceptional $E_9 = E_8^+$, only for $SU(5k)$, where k is an integer. This crucial emergence of ϕ from the E_8 lattice may be the key explanation of three other facts:

- McKay correspondence [84] between E_8 Lattice and double icosahedral group
- Correspondence between icosians and E_8 Lattice
- Discrete Hopf fibration $S^3 \to S^7 \to S^4$ [85]

We can now formulate our ontological theorem:

Theorem 4.1. E_8 *generates the golden ratio.*

Proof. The proof is given in equations 26 and 27. □

Proposition 4.2. *We propose the following ontological chain:* $\mathbb{B} \Rightarrow \mathbb{T} \Rightarrow E_8 \Rightarrow \phi \Rightarrow \mathbb{D} \Rightarrow QC \Rightarrow \mathfrak{h}_{92} \oplus \mathfrak{a}_7 \Rightarrow QSN$

- *The bits 0 and 1, or equivalently, the sphere S^0 emerges ab initio*
- *The trits, the integer, the matrix structure and the structured division algebra emerge from the bits, as described in Annex A (which also suggests a direct path to ϕ)*
- *The last division algebra is the octonions. Its integers, the Cayley integers, emerge as the E_8 lattice*
- *From theorem 4.1, the golden ratio ϕ emerges from the E_8 lattice*
- *Extending the integer by ϕ, our ring $\mathbb{D}$ of Dirichlet integers and its structural tensor products quantize all the needed algebras*
- *From the E_8 lattice, and in a simple way by using explicitly ϕ or in a more subtle way implicitly, emerges the Elser-Sloane quasicrystal*
- *Spacetime geometry emerges from this structure on which a gauge algebra is naturally inherited from the E_8 Lie group*
- *A 3D representation made of only regular tetrahedra forming a quasicrystal code made of Fibonacci chains, the QSN, emerges to support the geometrical reality*

Proof. We do not yet have a proof of this statement. However, each of the eight steps in the proposed ontological chain is mathematically sound and supported by evidence. The logical flow is strong. Accordingly, this is more than a conjecture. It is a deduced hypothesis. Readers are invited to think critically. This mathematical first-principles based emerging theory is incomplete but yet evidenced as the right direction via the physical evidence and equations presented herein. □

Quantum theory

An important result in foundational quantum mechanics is that of Hardy [86], which can be considered the best version of Bell's theorem [87] because of its simplicity and logical/mathematical rigor. While Bell's proof of the impossibility of Einstein, Podolsky and Rosen's view is based on statistical predictions and inequalities, Hardy's test is a test of non-locality without inequalities.

This test considers two qubits prepared in an entangled state, wherein each is sent in the opposite direction by a source in the middle of two detectors. One electron goes to the detector at left and the other to the right. The entanglement can be on polarization, spin components or momentum. The detectors have a switch between two positions that can be settled randomly and independently after the particles are emitted by the source and before it arrives at the detectors. We call the detector on the left L and on the right R. So they can have two positions $L1$, $L2$ and $R1$, $R2$. The choice for the switches are experiments with two possible outcomes, say ± 1, that depend on entangled particles. Just four of the possible experiments are needed to get the result of maximal non-locality, which we will present herein. These four experiments are fixed positions of the detectors to $L1$, $R1$ (let us call this experiment A); $L1$, $R2$ (experiment B), $L2$, $R1$ (experiment C) and $L2$, $R2$ (experiment D). Running these experiments many times, shows that, in experiment A, $L1$ and $R1$ being both -1 never occurs. In experiment B, with $L1$, $R2$, and experiment C, with $L2$ and $R1$, both being $+1$ never occurs. And in experiment D, $L2$ and $R2$ being both $+1$ occurs sometimes. According to local determinism, suppose that the atom going left has instructions to come out $+1$ if it encounters position L2. According to experiment C, if L2 gives $+1$, R1 can't give $+1$. So it must give -1. According to experiment A, if R1 gives 11, then L1 must give $+1$. And according to experiment B, if L1 gives $+1$, then R2 must give -1. This instruction set is the only one consistent with predictions A, B, and C and with L2 giving the result $+1$. It shows that, in situation D (for which the detectors are set to L2 and R2), whenever the left atom comes out $+1$, the right qubit must be -1. However, according to quantum mechanics, there is a probability for the right qubit to come out as $+1$. The exact probability P for $L2$ and $R2$ being both $+1$ is

$$P = \phi^{-5}. \tag{28}$$

This result is in strict accordance with both quantum mechanics experiment [88]. That is, the probability for the simplest quantity relationship, two, of the simplest fermions, electrons, in the simplest possible dynamic and geometric relationship (moving apart at equal velocities on the same line) is a probability of the golden ratio to the power of -5. This result is deep and important because, in some sense, this is the probability for maximal non-locality in nature.

Chaos theory: KAM theorem

The Kolmogorov, Arnold and Moser (KAM) theorem,[12] which is a result in classical mechanics of the study of dynamical systems. It is about the persistence of quasiperiodic motions under small perturbations. A dynamical

---

[12] There are many references in the context of chaos theory. For example, see [89, 90].

system can be expressed in configuration space in terms of action-angle coordinates consisting of action and angle variables in terms of a torus defined by its angle variables. Quasiperiodic orbits represent integral motion on a torus and, if it is integrable, there is a constant or invariant of the motion associated with the torus that leads to the term *invariant tori*. What happens to the invariant tori as the nonlinearity of the system increases? Say we have a system with Hamiltonian $H = H_o + \varepsilon H_1 + ...$, where H_o is the non-perturbed Hamiltonian and H_1 the non-linear perturbation allowing ε to mediate the force of the perturbation. For sufficiently small perturbations, virtually all tori are preserved.

Consider the frequency of motion around each angular variable of a torus. As a point moves, it rotates around the tube while revolving around the torus axis. If we take the ratio of these frequencies, we get a quantity called the winding number, $\sigma = \frac{\omega_1}{\omega_2}$. The KAM theorem shows that the tori most easily destroyed are those with rational winding numbers, while almost all other orbits (those with irrational winding numbers) are preserved. What happens if we increase the perturbation? The KAM theorem itself doesn't explicitly say, but derivative work does [91]. After the rational winding number tori go chaotic, the irrational tori eventually break up also, even though they are significantly more robust under perturbation. As the perturbation grows, more irrational tori go unstable. Most interestingly, the tori go unstable in order of the degree of irrationality of their the winding numbers. This is where golden ratio shows up as a physical limit. Mathematicians consider the golden ratio to be the most irrational of numbers because the rank of a number's irrationality is based on the speed of convergence of its continued fraction expression. The continued fraction expression for the golden ratio uses only the integer 1, making it the most "difficult" or slowest number to approximate with rational numbers. It approaches the limit slower than any other continued fraction expression.

It is at this golden ratio based winding number where physical vortices are most stable under increasing perturbations. This is the special limit related to how small denominators correspond to the growth of ε, allowing it to be minimal. It is the state where the conditions for the KAM theorem are most easily satisfied. In summary, golden ratio based gravitational, fluidic and other dynamical vortices are the most stable and therefore the most statistically probable physically abundant in nature.

Black hole physics

Black hole physics is a good laboratory for testing the limits of general relativity and quantum mechanics and for putting the two limits together in the study of quantum gravity theories. Accordingly, let us look for clues about Planck scale golden ratio physics in black hole equations. The first result is within the context of non-arbitrarily limiting and manipulating the equations of general relativity [92, 93]. This is a classic general relativity result with no relation to quantum mechanics. A rotating black hole can have a transition of phase between positive specific heat capacity and negative. The transition is based on the golden ratio when the ratio of angular momentum J and mass M is kept constant in the equation. In this case, ϕ is the point where a black hole's modified specific heat changes from positive to negative

$$\frac{M^4}{J^2} = \phi. \tag{29}$$

This non-arbitrary manipulation is analogous to the artificial but non-arbitrary setting of the velocity of the electron to zero in order to derive the electron rest mass. That is, although there is no such thing as an electron at rest, the non-arbitrary but non-physically realistic manipulation of the electron equation provides a deep fundamental understanding of a limit in nature. Let us now combine quantum mechanics with general relativity by considering the loop quantum gravity approach to quantum gravity [1]. Using this work done on the microstructure of spacetime, we compute black hole entropy. In a simplified framework, the isolated quantum horizons formulation of loop quantum gravity, we can derive the lower bound of black hole entropy [28]

$$e^{\frac{8\pi\hbar GS}{kA}} \geq \phi, \tag{30}$$

where S is the entropy, A the black hole area and k the Boltzmann constant. In section (3.1), we go further, using arguments from information theory to fix the loop quantum gravity parameter, which we can wright as

$$2^{\pi\gamma} = \phi. \tag{31}$$

5 Discussion

The (1) principle of least action and Noether's second theorem, (2) gauge symmetry, (3) general relativity and (4) the implication of spacetime discretization at the Planck scale strongly suggest an underlying new principle based on the efficiency of ratio, where, mathematically, the golden ratio is a special limit in the universe of all ratios and where, experimentally, it is observed in the fundamental physical equations and limits of nature.

Assuming that reality is discrete at the Planck scale, a logically satisfying explanation for the observed quantization of charge leads us to ask, "What is the most efficient organizing principle for how dynamical charge space is constructed?". A discrete reality is ultimately numerical, although not necessarily digital. It may be based on self-referential volumetric shape-numbers, as 3-simplex integers forming a quasicrystalline graph drawing based description of quantum gravity.

This multifaceted overall argument strongly implies that reality is code theoretic. The natural geometric languages that we know, such as DNA and other quasicrystals, are defined by the golden ratio, serving as experimental cases where the golden ratio is used by nature in geometric codes and inspiring the axiomatic theory that the physical universe behave with optimal digital efficiency by exploiting our foundational dimensionless constant ultimately residing at the Planck volume of space. This golden ratio based running constant starts at the quantum gravity scale and goes upwards fractally (all quasicrystals are fractal). The well-known self-similarity properties of the golden ratio and quasicrystals explain the observed fractality of nature at all scales [62, 95, 96, 97, 98, 99, 94], and provide a posteriori credit to the Dodecahedron-Icosahedron doctrine [100].

Note that, assuming the discreteness of spacetime at the Planck scale, we face a problem similar to one at the atomic scale: How do these building blocks self-organize to build more complex structures? How does order emerge? A crystalline structure would be a naive answer because crystals are not codes. Conversely, quasicrystalline structures are codes with non-local long range order. Syntactical freedom in their static and dynamic ordering rules provide the freedom necessary for quantum non-determinism and the ability for physicists to recover the gauge symmetry unification equations of particles and forces that express in the code due to the mapping of Lie algebras to the generating hyper-lattices from which quasicrystals are created.

6 Conclusion

We have presented a compelling idea that we can apply the results and tools from quantum information and quantum computation to a quantum spacetime code theoretic view using algebraic graph drawing formalism. We considered a DQW of a quantum particle on a quantum gravitational field and studied applications of related entanglement entropy. This memory time based entanglement entropy drives an entropic force, suggesting a unified picture of gravity and matter. Following this, we proposed a model for walker positions topologically encoded on a spin network, which can easily be re-expressed using twistors. This results in anomaly cancellation because the particles are no longer points but Planck scale voxels, as tetrahedral units of spacetime.

Later in this document, we presented a review of results that serve as compelling clues about an underlying organizing principle related to charge quantization and the golden ratio. The appearance of the golden ratio in black hole physics, the indication from the research of physics beyond the standard model and other evidences suggest there is quasicrystalline geometry at the unification scale and allow for a logical argument that nature has a limit of non-locality in a two particle quantum system. Again, an element of evidence for quasicrystalline structure at the quantum gravity scale. Due to the inherent fractal scale invariance of golden ratio based quasicrystals, this new efficiency and organizing principle shows up at other scales, such as the atomic scale, where, for example, quasicrystalline structure is present in the hydrogen atom and DNA. This unifying principle connecting the discussed points correlates nicely with derivative works from the KAM theorem, where it is trivially true that literally all systems in nature are both propagating and rotating simultaneously—forming propagating gravitational and electromagnetic vortices or extruded tori. The quasicrystalline phase is a special limit in thermal dynamics, where spatiotemporal degrees of freedom reach the maximum non- zero limit.

7   Acknowledgements

We also would like to acknowledge Carlos Castro Perelman for useful suggestions of references.

Appendix: Binary Matrices

We show that all n-dimensional structures used in relativity and quantum mechanics reduce to products of binary matrices. The quaternionic and complex structures, and the Pauli and Dirac algebra are illustrated below. The process [13] involves replacing i by a 2x2 anti-diagonal anti-symmetric "trit" (-1,0,1) matrix and then further reducing trit matrices by substituting -1 by 2x2 anti-diagonal symmetric non-null binary matrices.

A.1 Negative numbers and complex numbers

Algebraic structures emerge when a set of elements which, (along with operations) constitute our algebra, obey fundamental symmetries, eventually expressed as algebraic rules or constraints. For example, from the set of positive numbers, 0,1,2,3,4..., we define a new element m, outside of this set, satisfying $(m)^2 = 1$. So m is not 1. Indeed $m = -1$, and we have defined the negative unit, which we call -1. We extend our set and define $\mathbb{Z} = \mathbb{N}1 + \mathbb{N}m$ by using linear combinations of our old unit (1) and our new unit (m=-1) with coefficients in the old set $\mathbb{N}$. We can also use 2x2 matrices $1 = \begin{pmatrix} 1 & 0 \\ 0 & 1 \end{pmatrix}$ and $m = -1 = \begin{pmatrix} 0 & 1 \\ 1 & 0 \end{pmatrix}$ and check that $(m)^2 = 1$. Therefore the eigenvalues satisfy the same constraint. The complex numbers have a similar representation $z = a1 + bi = a \begin{pmatrix} 1 & 0 \\ 0 & 1 \end{pmatrix} + b \begin{pmatrix} 0 & -1 \\ 1 & 0 \end{pmatrix} = \begin{pmatrix} a & -b \\ b & a \end{pmatrix}$, where i satisfies $(i)^2 = -1$. The complex structure is the multiplicative group of the Gaussian integer units $\{1, -1, i, -i\}$. The trit to bit map μ from $\mathbb{T} = \{-1, 0, 1\}$ to $\mathcal{M}_2(\mathbb{F}_2)$ where $\mathbb{F}_2$ is the binary field $\{0,1\}$ is defined by $\mu(1) = \begin{pmatrix} 1 & 0 \\ 0 & 1 \end{pmatrix}, \mu(-1) = \begin{pmatrix} 0 & 1 \\ 1 & 0 \end{pmatrix}, \mu(0) = \begin{pmatrix} 0 & 0 \\ 0 & 0 \end{pmatrix}$ and can be applied to each matrix element obtained from the complex units by the map ν from the fourth roots of 1 to $\mathcal{M}_2(\mathbb{T})$ such that $\nu(i) = \begin{pmatrix} 0 & -1 \\ 1 & 0 \end{pmatrix}, \nu(-i) = \begin{pmatrix} 0 & 1 \\ -1 & 0 \end{pmatrix}, \nu(0) = \mu(0), \nu(1) = \mu(1), \nu(-1) = \mu(-1)$ to derive 4-dimensional binary matrices in $\mathcal{M}_4(\mathbb{F}_2)$ from the combined map $\rho = \mu \circ \nu$.

For example, $\rho(i) = \begin{pmatrix} \mu(0) & \mu(-1) \\ \mu(1) & \mu(0) \end{pmatrix} = \begin{pmatrix} 0 & 0 & 0 & 1 \\ 0 & 0 & 1 & 0 \\ 1 & 0 & 0 & 0 \\ 0 & 1 & 0 & 0 \end{pmatrix}.$

Note that in the two following subsections, the algebra of these matrices is not the standard algebra on $\mathcal{M}_n(\mathbb{R})$. It is also not the standard algebra on $\mathcal{M}_n(\mathbb{F}_2)$. The operations have to be computed in $\mathbb{R}$ and then projected to $\mathcal{M}_{\frac{n}{2}}(\{0_2, I_2, m\})$ by a process described in [13], where $\begin{pmatrix} 1 & 1 \\ 1 & 1 \end{pmatrix}$ projects to 0_2. In the alternative representation proposed on $\mathcal{M}_n(\mathbb{T})$, the projection is simply realized by the sign operator: $\forall\, x \in \mathbb{R}$, $\mathrm{sign}(x) =$
$$\begin{cases} -1 & \text{for} \quad x < 0 \\ 0 & \text{for} \quad x = 0 \\ 1 & \text{for} \quad x > 0 \end{cases}.$$

A.2 Quaternions and Pauli matrices

It was shown in [13] how to use 4D trit maps in $\mathcal{M}_4(\mathbb{T})$ or 8D bitmaps in $\mathcal{M}_8(\mathbb{F}_2)$ to represent quaternions and implement the quaternion group (finite group of degree 8), the D_4 root vectors group of degree 24 and the F_4 root vector group of degree 48. We repeat here the results for quaternions to deduce the Pauli matrices. We can define as κ the following map from the positive unit quaternions $\{1, i, j, k\}$ to $\mathcal{M}_4(\mathbb{T})$:

$$\kappa(i) = \begin{pmatrix} \nu(-i) & \nu(0) \\ \nu(0) & \nu(i) \end{pmatrix} = \begin{pmatrix} 0 & 1 & 0 & 0 \\ -1 & 0 & 0 & 0 \\ 0 & 0 & 0 & -1 \\ 0 & 0 & 1 & 0 \end{pmatrix}$$

$$\kappa(j) = \begin{pmatrix} \nu(0) & \nu(1) \\ -\nu(1) & \nu(0) \end{pmatrix} = \begin{pmatrix} 0 & 0 & 1 & 0 \\ 0 & 0 & 0 & 1 \\ -1 & 0 & 0 & 0 \\ 0 & -1 & 0 & 0 \end{pmatrix}$$

$\kappa(i)$ is named yin and $\kappa(j)$ is named yang in [13] because they generate the ring of 24 Hurwitz integers [101] with the modified trit matrix algebra projecting back the result by applying the sign function and its geometric dual. Naturally, the map κ is completed because the matrices satisfy Hamilton's quaternion relations $ijk = i^2 = j^2 = k^2 = -1$. $\kappa(k) = \kappa(i)\kappa(j) =$

$$\begin{pmatrix} \nu(0) & \nu(-i) \\ \nu(-i) & \nu(0) \end{pmatrix} = \begin{pmatrix} 0 & 0 & 0 & 1 \\ 0 & 0 & -1 & 0 \\ 0 & 1 & 0 & 0 \\ -1 & 0 & 0 & 0 \end{pmatrix}.$$

Quaternions are defined as binary matrices by applying μ to each element.

This generates the map $\lambda = \mu \circ \kappa$ from $\{1, i, j, k\}$ to $\mathcal{M}_8(\mathbb{F}_2)$ and

$$\lambda(i) = \begin{pmatrix} 0 & \mu(1) & 0 & 0 \\ \mu(-1) & 0 & 0 & 0 \\ 0 & 0 & 0 & \mu(-1) \\ 0 & 0 & \mu(1) & 0 \end{pmatrix} = \begin{pmatrix} 0&0&1&0&0&0&0&0 \\ 0&0&0&1&0&0&0&0 \\ 0&1&0&0&0&0&0&0 \\ 1&0&0&0&0&0&0&0 \\ 0&0&0&0&0&0&0&1 \\ 0&0&0&0&0&0&1&0 \\ 0&0&0&0&1&0&0&0 \\ 0&0&0&0&0&1&0&0 \end{pmatrix},$$

$$\lambda(j) = \begin{pmatrix} 0&0&0&0&1&0&0&0 \\ 0&0&0&0&0&1&0&0 \\ 0&0&0&0&0&0&1&0 \\ 0&0&0&0&0&0&0&1 \\ 0&1&0&0&0&0&0&0 \\ 1&0&0&0&0&0&0&0 \\ 0&0&0&1&0&0&0&0 \\ 0&0&1&0&0&0&0&0 \end{pmatrix}, \lambda(k) = \begin{pmatrix} 0&0&0&0&0&0&1&0 \\ 0&0&0&0&0&0&0&1 \\ 0&0&0&0&0&1&0&0 \\ 0&0&0&1&0&0&0&0 \\ 0&0&1&0&0&0&0&0 \\ 0&0&0&0&1&0&0&0 \\ 0&1&0&0&0&0&0&0 \\ 1&0&0&0&0&0&0&0 \end{pmatrix}. \quad (1)$$

Pauli matrices $\sigma_1 = \begin{pmatrix} 0 & 1 \\ 1 & 0 \end{pmatrix}$, $\sigma_2 = \begin{pmatrix} 0 & -i \\ i & 0 \end{pmatrix}$, $\sigma_3 = \begin{pmatrix} 1 & 0 \\ 0 & -1 \end{pmatrix}$ can be built by applying the map ρ to each of their elements $\{0, 1, -1, i, -i\}$:

$$\Lambda(\sigma_1) = \rho \begin{pmatrix} 0 & 1 \\ 1 & 0 \end{pmatrix} = \begin{pmatrix} \rho(0) & \rho(1) \\ \rho(1) & \rho(0) \end{pmatrix} = \begin{pmatrix} 0&0&0&0&1&0&0&0 \\ 0&0&0&0&0&1&0&0 \\ 0&0&0&0&0&0&1&0 \\ 0&0&0&0&0&0&0&1 \\ 1&0&0&0&0&0&0&0 \\ 0&1&0&0&0&0&0&0 \\ 0&0&1&0&0&0&0&0 \\ 0&0&0&1&0&0&0&0 \end{pmatrix},$$

$$\Lambda(\sigma_2) = \begin{pmatrix} 0&0&0&0&0&0&1&0 \\ 0&0&0&0&0&0&0&1 \\ 0&0&0&0&0&1&0&0 \\ 0&0&0&0&1&0&0&0 \\ 0&0&1&0&0&0&0&0 \\ 0&0&0&1&0&0&0&0 \\ 1&0&0&0&0&0&0&0 \\ 0&1&0&0&0&0&0&0 \end{pmatrix}, \Lambda(\sigma_3) = \begin{pmatrix} 1&0&0&0&0&0&0&0 \\ 0&1&0&0&0&0&0&0 \\ 0&0&1&0&0&0&0&0 \\ 0&0&0&1&0&0&0&0 \\ 0&0&0&0&0&1&0&0 \\ 0&0&0&0&1&0&0&0 \\ 0&0&0&0&0&0&0&1 \\ 0&0&0&0&0&0&1&0 \end{pmatrix}. \quad (2)$$

We define a new imaginary:

$$\imath = \rho(\sigma_1 \sigma_2 \sigma_3) = \rho \begin{pmatrix} i & 0 \\ 0 & i \end{pmatrix} = \begin{pmatrix} \rho(i) & \rho(0) \\ \rho(0) & \rho(i) \end{pmatrix} = \begin{pmatrix} 0&0&0&1&0&0&0&0 \\ 0&0&1&0&0&0&0&0 \\ 1&0&0&0&0&0&0&0 \\ 0&1&0&0&0&0&0&0 \\ 0&0&0&0&0&0&0&1 \\ 0&0&0&0&0&0&1&0 \\ 0&0&0&0&1&0&0&0 \\ 0&0&0&0&0&1&0&0 \end{pmatrix} \quad (3)$$

The quaternions $\lambda(i), \lambda(j), \lambda(k)$ and the Pauli matrices $\Lambda(\sigma_1), \Lambda(\sigma_2), \Lambda(\sigma_3)$ are related by: $\imath\lambda(i) = \Lambda(\sigma_3)$, $\lambda(j) = \imath\Lambda(\sigma_2)$, $\imath\lambda(k) = \Lambda(\sigma_1)$. Together with $1_8 = \rho(1_2)$ and $\imath$, they form the bi-quaternion units.

The Clifford algebra $\mathcal{Cl}_3$ also admits for basis in $\mathcal{M}_8(\mathbb{F}_2)$: $1_8, \Lambda(\sigma_1), \Lambda(\sigma_2)$, $\Lambda(\sigma_3), \lambda(i), \lambda(j), \lambda(k)$ and $\imath$.

They are illustrated below, where 0 is gray and 1 is white:

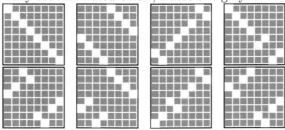

A.3 Dirac matrices

The 16 Dirac matrices $\gamma_{a,b}$, where $(a,b) \in \{0,1,2,3\}$, $\gamma_{0,0} = I_{16}$, $\gamma_{0,b} = I_2 \otimes \sigma_b$, $\gamma_{a,0} = \sigma_a \otimes I_2$ and $\gamma_{a,b} = \gamma_{a,0}\gamma_{0,b}$ are given in the below table, where 0 is gray and 1 is white:

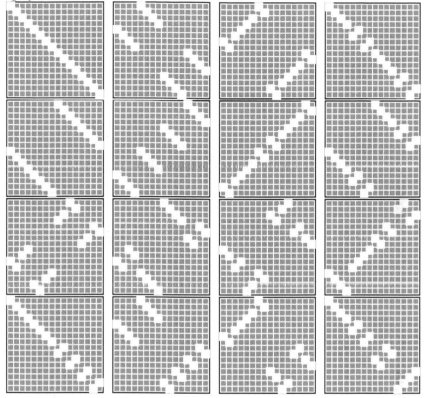

The same process can be extended based on $\lambda(i), \lambda(j), \lambda(k)$ or by combining the Dirac matrices with $\gamma_{0,\imath} = I_2 \otimes \imath$, $\gamma_{\imath,0} = \imath \otimes I_2$.

A.4 Binary matrix eigenvalues

These binary matrices have eigenvalues which are fourth roots of unity, the Gaussian integer units. We now systematically study all eigenvalues of n-dimensional binary matrices $\mathcal{M}_n(\mathbb{F}_2)$. We begin with n=2. There are 16 matrices. Their eigenvalues are, with $\phi = \frac{1+\sqrt{5}}{2}$ and $\varphi = \phi^{-1} = \frac{-1+\sqrt{5}}{2}$:

$\{0,0\}$, $\{1,0\}$, $\{0,0\}$, $\{1,0\}$, $\{0,0\}$, $\{1,0\}$,$\{-1,1\}$,$\{\phi,-\varphi\}$, $\{1,0\}$, $\{1,1\}$, $\{1,0\}$, $\{1,1\}$, $\{1,0\}$, $\{1,1\}$,$\{\phi,-\varphi\}$,$\{2,0\}$

Unexpectedly, the only non-integer eigenvalues of $\mathcal{M}_n(\mathbb{F}_2)$ are ϕ and $-\phi^{-1}$. This is one of the deepest mathematical reasons for our overall thesis that the unique qualities of golden ratio serve as an ultimately efficient fundamental constant of physics, expressing a ratio that is code-theoretically more powerful than 1.

From the eigenvalues of the 512 matrices of $\mathcal{M}_3(\mathbb{F}_2)$, the most probable are by decreasing order 0 (32%), 1 (28%), ϕ (6%), $-\phi^{-1}$ (6%), 2 (5%), -1 (5%)... From the eigenvalues of the 65,536 matrices of $\mathcal{M}_4(\mathbb{F}_2)$, the most probable are by decreasing order 0 (25%), 1 (18%), -1 (5%), 2 (4%), ϕ (4%), $-\phi^{-1}$ (4%), $1 - \sqrt{2}$ (1%), ... From the eigenvalues of the 33 554 432 matrices of $\mathcal{M}_5(\mathbb{F}_2)$, the most probable are by decreasing order 0 (19%), 1 (11%), -1 (4%), 2 (2%), ϕ (1%), $-\phi^{-1}$ (1%), $1 - \sqrt{2}$ (1%), ... They are the roots of 8,927 characteristic polynomials of degree 5. When the matrix dimension n is grows, the most probable eigenvalues have decreasing probabilities. ϕ always remains in the group of the most probable because its characteristic polynomial has small coefficients (1 or -1). At the other end, the less probable eigenvalue is n and appears only once, for the matrix made of only ones and polynomials $nx^{n-1} - x^n$. These most probable eigenvalues are the centers of exclusion circles with no other values and with radii proportional to their probabilities. We also see a smaller exclusion interval on the real line as illustrated in figure 13. This repelling behavior between eigenvalues is a known indicator of universality [102].

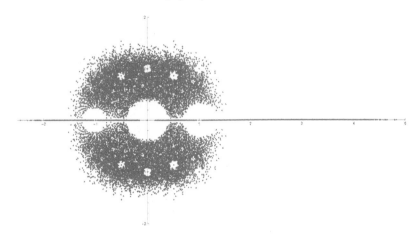

Figure 13: $\mathcal{M}_5(\mathbb{F}_2)$ spectrum

All of the above binary structure matrices representing the quaternions, biquaternions, Clifford algebra, Pauli matrices and Dirac matrices have $\mathcal{M}_n(\mathbb{F}_2)$ sub-blocks which are in $\{\mu(0), \mu(1) \text{ and } \mu(-1)\}$. By using the nilpotent sub-block $\epsilon = \begin{pmatrix} 0 & 1 \\ 0 & 0 \end{pmatrix}$, which has 0 as double-eigenvalue, we can implement dual integers. If we use the "golden" sub-block $\Phi = \begin{pmatrix} 0 & 1 \\ 1 & 1 \end{pmatrix}$, which has ϕ and $-\phi^{-1}$ as eigenvalues, we obtain an interesting structure. Note that in [77], the neutrino Majorana mass matrix $m_v = \begin{pmatrix} 0 & m & 0 \\ m & m & 0 \\ 0 & 0 & m_{atm} \end{pmatrix} = \left(\begin{array}{c|c} m\Phi & 0 \\ \hline 0 & m_{atm} \end{array} \right)$, which is block-diagonal with the sub-block Φ and therefore the mass eigenvalues make the neutrino-mixing angle $\theta_{12} = \arctan\left(\phi^{-1}\right)$.

If a, b, c and d are integers, $(aI_2 + b\Phi)(cI_2 + d\Phi) = (ac + bd)I_2 + (ad + bc + bd)\Phi)$ and the Dirichlet integer $(a + b\phi)$ can be implemented as $(a\mu(1) + b\Phi)$.

References

[1] C. Rovelli and F. Vidotto, "Covariant Loop Quantum Gravity," Cambridge University Press 1 edition, 2014.

[2] K. Irwin, "The Code-Theoretic Axiom", Full text: `https://www.researchgate.net/publication/314209684_The_Code_Theoretic_Axiom_The_Third_Ontology`. FQXI Essay Contest (2017), URL: `http://fqxi.org/community/forum/topic/2901`.

[3] R. Aschheim, "Hacking reality code". FQXI Essay Contest 2010-2011 - Is Reality Digital or Analog? (2011). URL: http://fqxi.org/community/forum/topic/929.

[4] J. A. Wheeler, "Hermann Weyl and the unity of knowledge", American Scientist 74 (1986): 366-375.

[5] J. A. Wheeler, "Information, physics, quantum: The search for links", In (W. Zurek, ed.) Complexity, Entropy, and the Physics of Information. Redwood City.CA: Addison-Wesley. Cited in DJ Chalmers,(1995) Facing up to the Hard Problem of Consciousness, Journal of Consciousness Studies 2.3 (1990): 200-19.

[6] M. Tegmark, "Is the theory of everything merely the ultimate ensemble theory?", Annals of Physics 270.1 (1998): 1-51.

[7] M. Tegmark, "The mathematical universe", Foundations of Physics 38.2 (2008): 101-150.

[8] D.B. Miller, E. Fredkin, "Two-state, Reversible, Universal Cellular Automata In Three Dimensions", Proceedings of the ACM Computing Frontiers Conference, Ischia, 2005.

[9] E. Fredkin, "Digital Mechanics", Physica D 45 (1990) 254-270.

[10] E. Fredkin, "An Introduction to Digital Philosophy", International Journal of Theoretical Physics, Vol. 42, No. 2 (2003) 189-247.

[11] S. Wolfram, "A New Kind Of Science", Wolfram Media, Inc., 2002.

[12] R. Aschheim, "SpinFoam with topologically encoded tetrad on trivalent spin networks," Loops11 talk, slide 10, 24 may 2011, 19h10, Madrid, arXiv:1212.5473.

[13] R. Aschheim, "Bitmaps for a Digital Theory of Everything," Presented at the 2008 Midwest NKS conference, Indiana University. Bloomington, IN, Oct. 31 - Nov 02, 2008. Available: http://www.cs.indiana.edu/~dgerman/2008midwestNKSconference/rasch.pdf.

[14] K. Irwin, "Toward the Unification of Physics and Number Theory". URL: https://www.researchgate.net/publication/314209738_Toward_the_Unification_of_Physics_and_Number_Theory.

[15] M. L. Senechal. "Quasicrystals and Geometry". Cambridge University Press, (1995).

[16] S. Hossenfelder, "Minimal length scale scenarios for quantum gravity", Living Rev. Rel. 16, 2 (2013)1203.6191[gr-qc].

[17] G. E. Gorelik, "Matvei Bronstein and quantum gravity: 70th an niversary of the unsolved problem", Phys. Usp., 48, 1039-1053 (2005).

[18] P. A. M. Dirac, "Quantised singularities in the electro magnetic field". Proc. Roy. Soc. London A 133, 60 (1931).

[19] C. Rovelli and F. Vidotto, "Single particle in quantum gravity and Braunstein-Ghosh-Severini entropy of a spin network," Phys. Rev. D 2010, 81, arXiv:0905.2983v2 [gr-qc].

[20] J. M. Garcia-Islas, "Entropic Motion in Loop Quantum Gravity," ARXIV gr-qc/1411.4383.

[21] G. Calcagni, D. Oriti and J. Thürigen, "Laplacians on discrete and quantum geometries," Class. Quantum Grav. 30 (2013) 125006, arXiv:1208.0354 [hep-th].

[22] D. Aharonov, A. Ambainis, J. Kempe, U. Vazirani, "Quantum Walks On Graphs," Proceedings of ACM Symposium on Theory of Computation (STOC'01), July 2001, p. 50-59, arXiv:quant-ph/0012090.

[23] M. Szegedy, "Quantum speed-up of Markov chain based algorithms," Proc. 45th IEEE Symposium on Foundations of Computer Science, pp. 32-41, 2004, arXiv:quant-ph/0401053.

[24] M. Szegedy, "Spectra of quantized walks and a rule", arXiv:quant-ph/0401053v1.

[25] B. Kollar, "Disorder and entropy rate in discrete time quantum walks," PhD thesis, 2014, Pecs, Hungary.

[26] A. Childs, "On the relationship between continuous and discrete time quantum walk," Commun.Math.Phys.,294,581, 2010, arXiv:0810.0312 [quant-ph].

[27] E. Verlinde, "On the Origin of Gravity and the Laws of Newton," JHEP 1104 (2011) 029, arXiv:1001.0785 [hep-th].

[28] C. Rovelli, "Black hole entropy from loop quantum gravity," Phys. Rev. Lett. **77**, 3288 (1996), arXiv:gr-qc/9603063.

[29] A. Ashtekar, J. C. Baez and K. Krasnov, "Quantum geometry of isolated horizons and black hole entropy," Adv. Theor. Math. Phys. **4**, 1 (2000), arXiv:gr-qc/0005126.

[30] M. Domagala and J. Lewandowski, "Black hole entropy from quantum geometry," Class. Quant. Grav. **21**, 5233 (2004), arXiv:gr-qc/0407051.

[31] A. Ghosh and P. Mitra, "An Improved lower bound on black hole entropy in the quantum geometry approach," Phys. Lett. B **616**, 114 (2005), arXiv:gr-qc/0411035.

[32] J. F. Barbero G. and A. Perez, "Quantum Geometry and Black Holes", arXiv:1501.02963 [gr-qc].

[33] S. Olsen, The Golden Section: Nature's Greatest Secret. New York, NY: Walker Books, 2006.

[34] M. J. Duff, L. B. Okun, G. Veneziano, "Trialogue on the number of fundamental constants". JHEP 0203 (2002) 023, arXiv:physics/0110060 [physics.class-ph].

[35] Y. M. Shnir, "Magnetic Monopoles". Springer, (2005).

[36] P. J. Mohr, B. N. Taylor and D. B. Newell. "CODATA Recommended Values of the Fundamental Physical Constants: 2010". http://physics.nist.gov/cuu/Constants/Preprints/lsa2010.pdf.

[37] G. Rosi, F. Sorrentino, L. Cacciapuoti, M. Prevedelli, G. M. Tino, "Precision Measurement of the Newtonian Gravitational Constant Using Cold Atoms". Nature 510, 518-521 (2014), arXiv:1412.7954 [physics.atom-ph].

[38] M. B. Green, J. H. Schwarz and E. Witten, "Superstring Theory", Vol. I and Vol. II, Cambridge University Press (1988).

[39] G. E. A. Matsas, V. Pleitez, A. Saa, D. A. T. Vanzella. "The number of dimensional fundamental constants". arXiv:0711.4276 [physics.class-ph] (2007).

[40] Private communication with Irvin Castro.

[41] V. Elser and N. J. A. Sloane. "A highly symmetric four-dimensional quasicrystal". J. Phys. A, 20:6161-6168, (1987).

[42] F. Fang and K. Irwin, "A Chiral Icosahedral QC and its Mapping to an E8 QC", Aperiodic2015 poster, arXiv:1511.07786 [math.MG].

[43] F. Fang, K. Irwin and R. Aschheim. "The Quasicrystalline Spin-Network - A Chiral Icosahedral Quasicrystal Derived from E 8", To appear.

[44] E. Witten. "Quest For Unification". arXiv:hep-ph/0207124.

[45] A. G. Lisi. "Lie Group Cosmology". arXiv:1506.08073 [gr-qc].

[46] F. Gursey, P. Ramond, P. Sikivie, Phys. Lett. B60 177 (1976); F. Gursey, M. Serdaroglu, Lett.Nuovo Cim.21:28, (1978).

[47] D. J. Gross, J. A. Harvey, E. Martinec and R. Rohm, Phys. Rev. Lett. 54 (1985), 502; Nucl. Phys. B 256 (1985), 253; ibid., B267 (1986), 75.

[48] P. Candelas, G. T. Horowitz, A. Strominger and E. Witten, Nucl. Phys. B 258 (1985), 46.

[49] Carlos Castro. "A Clifford algebra-based grand unification program of gravity and the Standard Model: a review study". Canadian Journal of Physics, 2014, 92(12): 1501-1527, 10.1139/cjp-2013-0686.

[50] I. Bars and M. Gunaydin, Phys. Rev. Lett 45 (1980) 859.

[51] N. Baaklini, Phys. Lett 91 B (1980) 376.

[52] S. Konshtein and E. Fradkin, Pisma Zh. Eksp. Teor. Fiz 42 (1980) 575.

[53] M. Koca, Phys. Lett 107 B (1981) 73.

[54] R. Coldea, D. A. Tennant, E. M. Wheeler, E. Wawrzynska, D. Prabhakaran, M. Telling, K. Habicht, P. Smeibidl, K. Kiefer. (2010), "Quantum Criticality in an Ising Chain: Experimental Evidence for Emergent E8 Symmetry", Science, 327 (5962): 177-180.

[55] M. Koca, "E8 lattice with icosians and Z5 symmetry". Journal of Physics A: Mathematical and General, 22(19):4125, 1989.

[56] L. Smolin, "The Trouble With Physics". Mariner Books; Reprint edition (September 4, 2007).

[57] R. Penrose, "The Road to Reality". Vintage; Reprint edition (January 9, 2007).

[58] M. Maggiore, Phys. Rev. D 49, 5182 (1994).

[59] A. N. Tawfik, A. M. Diab, "Review on Generalized Uncertainty Principle". ECTP-2015-01, WLCAPP-2015-01, arXiv:1509.02436 [physics.gen-ph].

[60] S. Liberati, L. Maccione, "Lorentz Violation: Motivation and new constraints". Ann.Rev.Nucl.Part.Sci.59:245-267,2009, arXiv:0906.0681 [astro-ph.HE].

[61] A. F. Ali, "Minimal Length in Quantum Gravity, Equivalence Principle and Holographic Entropy Bound". Class.Quant.Grav.28:065013,2011, arXiv:1101.4181 [hep-th].

[62] L. Nottale, "Scale Relativity And Fractal Space-Time". World Scientific Publishing Company; 1 edition (June 17, 2011).

[63] R. W. Numrich. "A metric space for computer programs and the principle of computational least action". The Journal of Supercomputing, 43(3):281-298, 2007.

[64] R. V. Moody and J. Patera. "Quasicrystals and icosians". Journal of Physics A: Mathematical and General, 26(12):2829, (1993).

[65] E. M. Barber. "Aperiodic structures in condensed matter". CRC Press, Taylor & Francis Group, (2009).

[66] J. D. Watson and F. H. Crick, "A Structure for Deoxyribose Nucleic Acid". Nature 171 (1953) 737-738.

[67] D. Levine and P. J. Steinhardt. "Quasicrystals. I. Definition and Structure." Physical Review B 34, no. 2 (July 15, 1986): 596-616. doi:10.1103/PhysRevB.34.596.

[68] M. Baake and U. Grimm. "Aperiodic Order". Cambridge University Press, (20130.

[69] J. Conway and N. Sloane. "Sphere Packing, Lattices and groups". Springer, (1998).

[70] J. Patera, editor. "Quasicrystals and discrete geometry", volume 10. American Mathematical Soc., (1998).

[71] E. Schrödinger. "What Is Life? The Physical Aspect of the Living Cell". Cambridge University Press, (1967).

[72] D. Shechtman, I. Blech, D. Gratias, and J. W. Cahn. "Metallic phase with long-range orientational order and no translational symmetry". Physical Review Letters, 53(20):1951, 1984.

[73] J. Brehm, W. Mullin, "Introduction to the Structure of Matter". John Wiley and Sons (1989).

[74] V. Petrusevski, Bulletin of the Chemists and Technologists of Macedonia, Vol. 25, No. 1, pp. 61-63 (2006).

[75] S. F. King and C. Luhn, "Neutrino Mass and Mixing with Discrete Symmetry". Rept. Prog. Phys. **76**, 056201 (2013) [arXiv:1301.1340 [hep-ph]].

[76] A. Datta, F. S. Ling and P. Ramond, "Correlated hierarchy, Dirac masses and large mixing angles". Nucl. Phys. B **671**, 383 (2003) [hep-ph/0306002].

[77] Y. Kajiyama, M. Raidal and A. Strumia, "The Golden ratio prediction for the solar neutrino mixing". Phys. Rev. D **76**, 117301 (2007) [arXiv:0705.4559 [hep-ph]].

[78] L. L. Everett and A. J. Stuart, "Icosahedral (A(5)) Family Symmetry and the Golden Ratio Prediction for Solar Neutrino Mixing". Phys. Rev. D **79**, 085005 (2009) [arXiv:0812.1057 [hep-ph]].

[79] I. de Medeiros Varzielas and L. Lavoura, "Golden ratio lepton mixing and nonzero reactor angle with A_5". J. Phys. G **41**, 055005 (2014) [arXiv:1312.0215 [hep-ph]].

[80] F. Feruglio and A. Paris, "The Golden Ratio Prediction for the Solar Angle from a Natural Model with A5 Flavour Symmetry". JHEP **1103**, 101 (2011). [arXiv:1101.0393 [hep-ph]].

[81] Q. Duret, B. Machet, "The neighborhood of the Standard Model: mixing angles and quark-lepton complementarity for three generations of non-degenerate coupled fermions". [arXiv:0705.1237 [hep-ph]].

[82] H. Minakata and A. Y. Smirnov, "Neutrino mixing and quark-lepton complementarity". Phys. Rev. D **70**, 073009 (2004). [hep-ph/0405088].

[83] S. F. King, "Tri-bimaximal-Cabibbo Mixing". Phys. Lett. B **718**, 136 (2012). [arXiv:1205.0506 [hep-ph]].

[84] S.P. Sirag, ADEX Theory: How the ADE Coxeter Graphs Unify Mathematics and Physics. Series on Knots and Everything, vol. 57. New Jersey: World Scientific, 2016.

[85] J. F. Sadoc and R. Mosseri, "The E8 lattice and quasicrystals: geometry, number theory and quasicrystals". Journal of Physics A: Mathematical and General vol. 26, 1789 (1993);

[86] L. Hardy, 1993. "Nonlocality for two particles without inequalities for almost all entangled states". Physical Review Letters, 71(11), p.1665.

[87] N. D. Mermin, "The Best Version of Bell's Theorem" N.Y. Acad. Sci.755, 616 (1995).

[88] A. G. White, D. F.V. James, P. H. Eberhard, and P. G. Kwiat, "Non-maximally entangled states: production, characterization, and utilization" Physical Review Letters, 83 (1999) 3103-3107.

[89] I. C. Percival, "Chaos in Hamiltonian Systems" Proc. R. Soc. Lond. A 1987 413 , 131-143.

[90] M. Toda, T. Komatsuzaki, T. Konishi, R. S. Berry, S. A. Rice, "Geometric Structures of Phase Space in Multidimensional Chaos: Applications to Chemical Reaction Dynamics in Complex Systems" Volume 130, 2005 John Wiley and Sons, Inc.

[91] J. M. Greene, "A method for determining a stochastic transition" J. Math. Phys. 20 1183 (1979).

[92] P. C. W. Davies, "Thermodynamic phase transitions of Kerr-Newman black holes in de Sitter space", Classical and Quantum Gravity 6 (1989), 1909-1914.

[93] J. C. Baez, `https://johncarlosbaez.wordpress.com/2013/02/28/` `black-holes-and-the-golden-ratio/`.

[94] C. G. Langton, "Computation at the edge of chaos: Phase transitions and emergent computation". Physica D: Nonlinear Phenomena, vol. 42, no. 1, pp. 12-37, (1990).

[95] M. Niedermaier and M. Reuter, "The asymptotic safety scenario in quantum gravity". Living Rev. Relat. 9, 5 (2006).

[96] M. Reuter and F. Saueressig, "Asymptotic safety, fractals, and cosmology". Lect. Notes Phys. 863, 185 (2013) [arXiv:1205.5431].

[97] J. Ambjørn, A. Görlich, J. Jurkiewicz, and R. Loll, "Nonperturbative quantum gravity". Phys. Rep. 519, 127 (2012) [arXiv:1203.3591].

[98] G. Calcagni, "Discrete to continuum transition in multifractal spacetimes". Phys. Rev. D 84, 061501(R) (2011) [arXiv:1106.0295].

[99] M. Arzano, G. Calcagni, D. Oriti, and M. Scalisi, "Fractional and noncommutative spacetimes". Phys. Rev. D 84, 125002 (2011) [arXiv:1107.5308].

[100] A. Stakhov and S. Olsen. The Mathematics of Harmony: From Euclid to Contemporary Mathematics and Computer Science. Vol. 22. Series on Knots and Everything. WORLD SCIENTIFIC, 2009. http://www.worldscientific.com/worldscibooks/10.1142/6635.

[101] A. Hurwitz. "Vorlesungen Uber die Zahlentheorie der Quaternionen". Published: Berlin: J. Springer, IV u. 74 S. gr. 8° (1919).

[102] L. Erdös, "Universality of Wigner Random Matrices: A Survey of Recent Results." Russian Mathematical Surveys 66, no. 3 (June 30, 2011): 507-626. doi:10.1070/RM2011v066n03ABEH004749.

9   Spontaneous symmetry breaking and the Unruh effect

Antonio Dobado

Abstract   In this work we consider the ontological status of the Unruh effect. Is it just a formal mathematical result? Or the temperature detected by an accelerating observer can lead to real physical effects such as phase transition. In order to clarify this issue we use the Thermalization Theorem to explore the possibility of having a restoration of the symmetry in a system with spontaneous symmetry breaking of an internal continuous symmetry as seen by an accelerating observer. We conclude that the Unruh effect is an ontic effect, rather than an epistemic one, giving rise, in the particular example considered here, to a phase transition (symmetry restoration) in the region close to the accelerating observer horizon.

1   Introduction

Trying to understand better Hawking radiation [1], Unruh did an amazing discovery in 1976 [2] (see [3] for a very complete review). He realized that an observer moving through the Minkowski vacuum with a constant acceleration a will detect a thermal bath at temperature:

$$T = \frac{a\hbar}{2\pi c k_B}. \tag{1}$$

This result was first obtained for free bosonic quantum fields but later it was extended to interacting fields giving rise to the so called Thermalization Theorem [4]. The relevance of the above formula is based, among other things, on the fact that it relates Quantum Mechanics, Relativity and Statistical Physics because it contains the Planck constant $\hbar$, the speed of light c and the Boltzmann constant k_B (in the following we will use natural units with $c = \hbar = k_B = 1$).

There are different approaches to the Unruh effect. The first one is based

A. S. Stefanov, M. Giovanelli (Eds), *General Relativity 1916 - 2016. Selected peer-reviewed papers presented at the Fourth International Conference on the Nature and Ontology of Spacetime, dedicated to the 100th anniversary of the publication of General Relativity, 30 May - 2 June 2016, Golden Sands, Varna, Bulgaria* (Minkowski Institute Press, Montreal 2017). ISBN 978-1-927763-46-9 (softcover), ISBN 978-1-927763-47-6 (ebook).

on Bogolyubov transformations, and it was the approach used by the pioners of field quantization on Rindler space [5, 6, 7, 8]. There is also the operational approach based in the concept of Unruh-DeWitt detector where one studies the response of accelerating detectors to the quantum fluctuations of the fields. Also it is possible to use operator algebra in the context of Modular Theory where the concept of KMS (Kubo-Martin-Schwinger [9], [10]) states plays a dominant role (see [11] for detailed review). This is possibly the most abstract approach and the most far away from the physical interpretation of the phenomenon. On the opposite side one can consider the *experimental* approach based in analogue systems [12] suggested by Unruh himself [13], as for example by studying the behavior of subsonic-ultrasonic interfaces in Bose-Einstein condensates [14]. Moreover Bell and Leinaas have suggested the possibility of observing the Unruh effect by measuring the polarization of electrons in storage rings [15].

Finally we have the so-called Thermalization Theorem. It was introduced by Lee [4] and it is based in a path integral approach to Quantum Field Theory (QFT) in curved spacetime. This approach is the most general since it incorporates many elements of the previous ones. In addition it can be applied to any kind of fields like scalars, gauge or fermions and most importantly, to interacting systems. It also offers a picture of the Unruh effect as an instance of the Einstein-Podolsky-Rosen Gedanken experiment [16].

In this work we want to explore the possibility of the Unruh effect for producing non-trivial thermal effects such as phase transitions. For this reason we will be using the Thermalization Theorem approach appropriate for interacting QFT. In particular we will study the symmetry breaking restoration produced by acceleration in the $SO(N + 1)$ Linear Sigma Model (LSM). At zero temperature and Minkowski space this model can (for appropriate values of the parameters) feature spontaneous symmetry breaking (SSB) from $SO(N+1)$ down to $SO(N)$. The model is renormalizable and, in addition, it can be solved in a non-perturbative way for large N in a particular limit. Also it has a thermal second order $SO(N+1)$ symmetry restoration at a temperature $T_c = 2v\sqrt{3/N}$ in the large N limit, with v being the vacuum expectation value (VEV). By using the Thermalization Theorem we will show that an accelerating observer will indeed observe a restoration of the symmetry in this model at some critical acceleration:

$$a_c = 4\pi v \sqrt{\frac{3}{N}}. \tag{2}$$

Moreover the accelerating observer will see a different value of the VEV at different distances from her horizon so that the restoration of symmetry is produced in the region close to this horizon. Some previous related results have been obtained for the Nambu-Jona-Lasinio model in [17] and [18] and in [19] for the $\lambda\Phi^4$ theory at the one-loop level.

2 Comoving coordinates for Rindler space

When dealing with accelerating observers (or detectors) in Minkowski space it is very useful to use Rindler and comoving coordinates. In the four-dimensional Minkowski space M we can introduce Cartesian inertial coordinates $X^\mu = (T, X, Y, Z)$ with metric:

$$ds^2 = dT^2 - dX^2 - dY^2 - dZ^2 \tag{1}$$

We define also $X_\perp = (Y, Z)$ so that:

$$ds^2 = dT^2 - dX^2 - dX_\perp^2 \tag{2}$$

in an obvious notation. Clearly these coordinates move along the whole real axis $T, X, Y, Z \in (-\infty, \infty)$. Next we can introduce Rindler coordinates as follows:

$$
\begin{aligned}
T &= \rho \sinh \eta \\
X &= \rho \cosh \eta
\end{aligned}
\tag{3}
$$

where $\rho \in (0, \infty)$ and $\eta \in (-\infty, \infty)$. Thus these coordinates are covering only the region $X > | T |$, the so called $\mathcal{R}$ wedge. It is also possible to introduce the complementary coordinates ρ' and η' as:

$$
\begin{aligned}
T &= \rho' \sinh \eta' \\
X &= -\rho' \cosh \eta'
\end{aligned}
\tag{4}
$$

covering the left wedge $\mathcal{L}$ where $-X > | T |$. In the case of the $\mathcal{R}$ region the metric reads:

$$ds^2 = \rho^2 d\eta^2 - d\rho^2 - dX_\perp^2. \tag{5}$$

The other two regions are the origin past $\mathcal{P}$ ($T < - | X |$) and future $\mathcal{F}$ ($T > | X |$).

In Minkowski space, an uniformly accelerating observer in the X direction will follow a world line like:

$$
\begin{aligned}
T(\tau) &= \frac{1}{a} \sinh (a\tau) \\
X(\tau) &= \frac{1}{a} \cosh (a\tau)
\end{aligned}
\tag{6}
$$

where we assume $X_\perp$ to be constant and $a > 0$ and τ are the proper acceleration and the proper time respectively. The same world line is described in Rindler coordinates by the simple equations:

$$
\begin{aligned}
\rho &= \frac{1}{a} \\
\eta &= a\tau.
\end{aligned}
\tag{7}
$$

Therefore Rindler coordinates correspond to a network of observers with different proper constant acceleration $a = 1/\rho$ and having a clock measuring their proper times in a units. Those observers have a past and a future horizon at $X = -T$ and $X = T$ respectively that they find in the infinite remote past or future (in proper time) or also in the limit $\rho \to 0$ (infinite acceleration).

In the following it will be very interesting to introduce a new system of coordinates on the manifold $\mathcal{R}$, i.e. $x^\mu = (t, x, y, z)$ defined as:

$$
\begin{aligned}
T &= \frac{1}{a} e^{ax} \sinh(at) \\
X &= \frac{1}{a} e^{ax} \cosh(at) \\
Y &= y \\
Z &= z.
\end{aligned}
\tag{8}
$$

These are just the comoving coordinates of a non-rotating accelerating observer with constant acceleration a in the X direction. Note that $t, x, y, z \in (-\infty, \infty)$ and one has $\rho = e^{ax}/a$ and $\eta = at$. Thus a point with fixed x coordinate is having an acceleration:

$$
a(x) = a e^{-ax}
\tag{9}
$$

so that $a(0) = a$ but $a(x)$ goes to infinity when x goes to $-\infty$ (the horizon) and it goes to zero when x goes to ∞. In these comoving coordinates the metric has the simple form:

$$
ds^2 = e^{2ax}(dt^2 - dx^2) - dx_\perp^2
\tag{10}
$$

where $x_\perp = (y, z)$.

Alternatively it is possible to define the coordinates $x'^\mu = (t', x', y', z')$ as:

$$
\begin{aligned}
T &= \frac{1}{a} e^{ax'} \sinh(at') \\
X &= -\frac{1}{a} e^{ax'} \cosh(at') \\
Y &= y' \\
Z &= z'.
\end{aligned}
\tag{11}
$$

These coordinates correspond to a comoving observer having a constant acceleration a along the negative X direction. As it is easy to show, they can be used to cover the $\mathcal{L}$ wedge. In terms of these coordinates the metric has exactly the same form as in the previous case ($\mathcal{R}$ wedge). A very important remark when comparing Rindler coordinates in Eq.(3) with comoving coordinates in Eq.(8) is that Rindler coordinates do not show any dimensional parameter or physical scale. In this sense they are similar to Minkowski coordinates. However comoving coordinates refer to a particular observer with

164

acceleration a and thus they depend on this physical scale. This fact will become relevant later in this work.

3   The Thermalization Theorem

The accelerating observer can only feel directly the Minkowski vacuum fluctuations inside $\mathcal{R}$. However those fluctuations are entangled with the ones corresponding to the left Rindler region $\mathcal{L}$ ($X < - \mid T \mid$) as in a kind of Einstein, Podolsky and Rosen setting. The result is that she sees the Minkowski vacuum as a mixed state described by a density matrix ρ_R which, according to the Thermalization Theorem [4], can be written in terms of the Rindler Hamiltonian $\hat{H}_R$ (the generator of the t time translations) as:

$$\hat{\rho}_R = \frac{e^{-2\pi\hat{H}_R/a}}{Tr e^{-2\pi\hat{H}_R/a}}. \tag{1}$$

Thus the expectation value of any operator $\hat{A}_R$ defined on the Hilbert space $\mathcal{H}_R$ corresponding to the region $\mathcal{R}$ in the Minkowski vacuum $\mid \Omega_M >$ is given by:

$$< \Omega_M \mid \hat{A}_R \mid \Omega_M >= Tr\hat{\rho}_R\hat{A}_R. \tag{2}$$

This result can be seen as the one found in a thermal ensemble at temperature $T = a/2\pi$ (in natural units) and it can be understood as a very precise formulation of the Unruh effect.

In any case one can of course wonder about the ontological status of this effect. Is the above result just formal or does it truly represent a thermal effect? In more prosaic terms: Could it be possible to cook a steak by accelerating it? More technically speaking: Can the Unruh effect give rise to phase transitions?

4   The spontaneously broken $SO(N+1)$ Linear Sigma Model

In order to explore this issue we have considered a model featuring a spontaneous symmetry breaking, namely the well known $SO(N+1)$ Linear Sigma Model (LSM). This model is defined in Minkowski space by the Lagrangian:

$$\mathcal{L} = \frac{1}{2}\partial_\mu\Phi^T\partial^\mu\Phi - V\left(\Phi^T\Phi\right) + J\sigma \tag{1}$$

where the multiplet $\Phi^T = (\bar{\pi}, \sigma)$ contains $N+1$ real scalar fields ($\bar{\pi}$ is an N component scalar multiplet). The potential is given by:

$$V\left(\Phi^T\Phi\right) = -\mu^2\Phi^T\Phi + \lambda\left(\Phi^T\Phi\right)^2 \tag{2}$$

where λ is positive in order to have a potential bounded from below and μ^2 is positive in order to produce a spontaneous symmetry breaking (SSB). When the external field is turned off $(J = 0)$, the SSB pattern is $SO(N + 1) \rightarrow SO(N)$ and N Nambu-Goldstone bosons appear in the spectrum.

At the tree level and $a = 0$ the low-energy dynamics is controlled by the broken phase where:

$$< \Omega_M \mid \hat{\pi}^a \mid \Omega_M >= 0; \quad < \Omega_M \mid \hat{\sigma} \mid \Omega_M >= v. \tag{3}$$

and $v^2 = \mu^2/2\lambda = NF^2$ with F being N independent. Then the relevant degrees of freedom are the $\hat{\pi}$ fields, which correspond to the Nambu-Goldstone bosons (pions). Fluctuations along the σ direction correspond to the Higgs, the massive mode which is relevant at higher energies or temperatures.

According to the Thermalization Theorem an accelerating observer will see the system as a canonical ensemble described by the partition function given by:

$$Z_R(a) = Tr e^{-\frac{2\pi}{a}\hat{H}_R} = \int [d\Phi] \exp\left(-S_{RE}[\Phi]\right) , \tag{4}$$

with the thermal like periodic boundary conditions in Euclidean signature:

$$\Phi(\bar{x}, 0) = \Phi(\bar{x}, 2\pi/a) \tag{5}$$

and also

$$\Phi(\mid \bar{x} \mid = \infty, t_E)^T \Phi(\mid \bar{x} \mid = \infty, t_E) = \sigma^2(\mid \bar{x} \mid = \infty, t_E) = v^2, \tag{6}$$

where t_E is the Euclidean comoving time. In comoving coordinates the Euclidean action $S_{RE}[\Phi]$ defined on $\mathcal{R}$ is:

$$
\begin{aligned}
S_{RE}[\bar{\pi}, \sigma] = & \frac{1}{2} \int d^4x((\partial_t\bar{\pi})^2 + (\partial_t\sigma)^2 + (\partial_x\bar{\pi})^2 + (\partial_x\sigma)^2 \\
& + \sqrt{g}[(\nabla_\perp\bar{\pi})^2 + (\nabla_\perp\sigma)^2 + 2\lambda(\bar{\pi}^2 + \sigma^2)^2 - 2\mu^2(\bar{\pi}^2 + \sigma^2)])
\end{aligned}
\tag{7}
$$

with

$$\int d^4x = \int_0^{2\pi/a} dt_E \int_{-\infty}^{\infty} dx \int_{-\infty}^{\infty} dy \int_{-\infty}^{\infty} dz \tag{8}$$

and $\sqrt{g} = e^{2ax}$. In order to compute the partition function in the large N limit a standard technique consists in introducing an auxiliary scalar field ϕ as follows. The quartic term appearing in the above partition function is:

$$\exp\left(-\int d^4x\sqrt{g}\lambda(\bar{\pi}^2 + \sigma^2)^2\right). \tag{9}$$

This term can be taken into account just by introducing in the action:

$$-\frac{1}{2}\int d^4x\sqrt{g}\left(N\phi^2 - \sqrt{8\lambda N}\phi(\bar{\pi}^2 + \sigma^2)\right) \tag{10}$$

and performing an additional $[d\phi]$ functional integration after the integrations on the $\bar{\pi}$ and σ fields.

The (algebraic) Euler-Lagrange equation for ϕ simply gives:

$$\phi^2 = \frac{2\lambda}{N}(\bar{\pi}^2 + \sigma^2)^2. \tag{11}$$

Therefore the partition function can then be written as:

$$Z_R(a) = \int [d\phi][d\sigma][d\bar{\pi}] \exp\left(-S_{RE}[\bar{\pi}, \sigma, \phi]\right). \tag{12}$$

Notice that now all the interactions are mediated by the new auxiliary field ϕ. In terms of the $\bar{\pi}$, σ and ϕ fields the action reads:

$$
\begin{aligned}
S_{RE}[\bar{\pi}, \sigma, \phi] &= \int d^4x \sqrt{g} \, [\frac{1}{2}\pi^a \left(-\Box_E - 2\mu^2 + \sqrt{8\lambda N}\phi\right)\pi^a \\
&+ \frac{1}{2}\sigma \left(-\Box_E - 2\mu^2 + \sqrt{8\lambda N}\phi\right)\sigma - \frac{1}{2}N\phi^2].
\end{aligned}
\tag{13}
$$

At this point it is convenient to introduce the new field:

$$\chi = 4\lambda(\bar{\pi}^2 + \sigma^2 - v^2) = \phi\sqrt{8\lambda N} - 2\mu^2. \tag{14}$$

Then the above action becomes:

$$
\begin{aligned}
S_{RE}[\bar{\pi}, \sigma, \chi] &= \int d^4x \sqrt{g}[\frac{1}{2}\pi^a \left(-\Box_E + \chi\right)\pi^a + \frac{1}{2}\sigma \left(-\Box_E\right)\sigma \\
&+ \frac{1}{2}(\sigma^2 - v^2)\chi - \frac{\chi^2}{16\lambda} - \lambda v^4].
\end{aligned}
\tag{15}
$$

By performing a standard Gaussian integration of the pion fields we get:

$$\int [d\bar{\pi}] \exp\left(-\frac{1}{2}\int d^4x \, \sqrt{g}\pi^a \left[-\Box_E + \chi\right]\pi^a\right) = \exp\left(-\Delta\Gamma[\chi]\right). \tag{16}$$

where:

$$\Delta\Gamma[\chi] = \frac{N}{2}Tr\log\frac{-\Box_E + \chi}{-\Box_E}. \tag{17}$$

Thus we have:

$$Z_R(a) = \int [d\chi][d\sigma]e^{-\Gamma_R[\sigma,\chi]} \tag{18}$$

where the effective action in the exponent is:

$$
\begin{aligned}
\Gamma_R[\sigma, \chi] &= \int d^4x \sqrt{g}[\frac{1}{2}\sigma \left(-\Box_E\right)\sigma + \frac{1}{2}\left(\sigma^2 - v^2\right)\chi \\
&- \frac{\chi^2}{16\lambda} - \lambda v^4 + \frac{N}{2}\log\frac{-\Box_E + \chi}{-\Box_E}]
\end{aligned}
\tag{19}
$$

At the leading order in the large N expansion this is all we need since we can expand the effective action around some given field configuration $\bar{\sigma}$ and $\bar{\chi}$ as:

$$\Gamma_R[\sigma,\chi] = \Gamma_R[\bar{\sigma},\bar{\chi}] + \int d^4x\sqrt{g}\frac{\delta\Gamma_R}{\delta\sigma(x)}\delta\sigma(x) + \int d^4x\sqrt{g}\frac{\delta\Gamma_R}{\delta\chi(x)}\delta\chi(x) + ... \quad (20)$$

Now we can choose $\bar{\sigma}$ and $\bar{\chi}$ as the solutions of:

$$\frac{\delta\Gamma_R}{\delta\sigma(x)} = -\Box_E\sigma + \chi\sigma = 0 \quad\quad (21)$$

$$\frac{\delta\Gamma_R}{\delta\chi(x)} = \frac{1}{2}\left(\sigma^2 - v^2\right) - \frac{\chi}{8\lambda} + \frac{\delta}{\delta\chi(x)}\frac{N}{2}Tr\log\frac{-\Box_E + \chi}{-\Box_E} = 0 \quad (22)$$

so that, by using the saddle point approximation:

$$Z_R(a) = e^{-\Gamma[\bar{\sigma},\bar{\chi}]} + O(1/\sqrt{N}), \quad\quad (23)$$

where we have taken into account that $\Gamma[\sigma,\chi]$ is order N. This large N approximation must be understood as $N \to \infty$ with $\lambda \to 0$ while keeping λN finite. Then, in this limit we have:

$$\bar{\sigma}(x) = <\Omega_M\mid\hat{\sigma}(x)\mid\Omega_M> \quad\quad (24)$$
$$\bar{\sigma}^2(x) = <\Omega_M\mid(\hat{\sigma}(x))^2\mid\Omega_M>.$$

5 The VEV in the Minkoski vacuum as seen in the comoving frame

In order to solve the above equations to obtain $\bar{\sigma}$ and $\bar{\chi}$ we first realize that, in the large ρ limit, and keeping $ax << 1$, the accelerating observer goes into the Minkowski inertial frame which in turns means that $\bar{\sigma}$ goes to v and $\bar{\chi}$ goes to zero. In fact those are the boundary conditions needed for the the applicability of the thermalization theorem to the system considered here. Therefore it makes sense trying to solve the equations in the $\chi = 0$ and $ax << 1$ regime. In that case we have:

$$0 = \Box_E\sigma \quad\quad (1)$$

$$0 = \sigma^2 - v^2 + \frac{N}{2\pi^3}\int_0^\infty d\Omega\frac{\Omega\pi}{2\rho^2\tanh(\Omega\pi)}. \quad\quad (2)$$

By writing Ω as ω/a and using $\rho a = 1 + ax + ...$ we find, up to order ax:

$$\sigma^2 = v^2 - \frac{N}{4\pi^2}(1-2ax)\int_0^\infty d\omega\omega\left(1 + \frac{2}{e^{\frac{2\pi}{a}\omega} - 1}\right), \quad\quad (3)$$

where the first divergent integral requires regularization and renormalization. This can be done by using a x dependent ultraviolet cutoff Λe^{-ax} to compute the divergent integral and performing the renormalization of the v parameter:

$$v^2 \rightarrow v^2 - N\frac{\Lambda^2}{2(2\pi)^2}(1 - 2ax + ...). \tag{4}$$

This renormalization is compatible with the limit $a = 0$ (Minkowski inertial coordinates) and with the red/blue shift detected by the accelerating observer when recieving a signal emmitted at the point x. A similar result can be obtained by using dimensional renormalization. In any case we have:

$$\sigma^2 = v^2 - \frac{N}{2\pi^2}(1 - 2ax)\int_0^\infty d\omega\omega\frac{1}{e^{\frac{2\pi}{a}\omega} - 1} + O((ax)^2). \tag{5}$$

By performing the ω integration, the Minkowski VEV of the $\hat{\sigma}^2(x)$ comoving operator is given in the $ax << 1$ regime by:

$$\bar{\sigma}^2(x) = <\Omega_M \mid (\hat{\sigma}(x))^2 \mid \Omega_M> = v^2\left(1 - \frac{a^2 N}{12(2\pi)^2v^2}(1 - 2ax)\right). \tag{6}$$

By introducing the critical acceleration:

$$a_c^2 = 3(4\pi)^2\frac{v^2}{N} \tag{7}$$

we have:

$$\bar{\sigma}^2(x) = v^2\left(1 - \frac{a^2}{a_c^2} + 2\frac{a^3}{a_c^2}x + ...\right). \tag{8}$$

Notice that at this order this is also a solution of Eq. (21). Therefore, at the origin of the accelerating frame ($x = 0$ or $\rho = 1/a$), the squared VEV of the $\hat{\sigma}$ field is given by:

$$\bar{\sigma}^2(0) = <\Omega_M \mid (\hat{\sigma}(0))^2 \mid \Omega_M> = v^2\left(1 - \frac{a^2}{a_c^2}\right) \tag{9}$$

for $0 \leq a \leq a_c$ and clearly:

$$<\Omega_M \mid (\hat{\sigma}(0))^2 \mid \Omega_M> = 0 \tag{10}$$

for $a > a_c$. This is exactly the thermal behavior of the LSM in the large N limit with a/a_c playing the role of T/T_c (as seen by a inertial observer). It corresponds to a second order phase transition at the critical acceleration a_c where the original spontaneously broken symmetry is restored for the accelerating observer.

Now let us consider a different accelerating observer at Rindler coordinate $\rho' = 1/a'$. This observer will find a similar result just changing a by a'. From the point of view of the first observer the second observer is located at some point x' given by:

$$\rho' = \frac{1}{a'} = \frac{1}{a}e^{ax'} \tag{11}$$

i.e. the acceleration of the second observer is $a' = ae^{-ax'}$. In this way it is immediate to find the position dependent result for the squared VEV of the σ field which, in comoving coordinates, is given by:

$$\bar{\sigma}^2(x) = <\Omega_M \mid (\hat{\sigma}(x))^2 \mid \Omega_M> = v^2 \left(1 - \frac{a^2}{a_c^2} e^{-2ax} \right) \tag{12}$$

or, in Rindler coordinates, by:

$$\bar{\sigma}^2(\rho) = v^2 \left(1 - \frac{1}{a_c^2 \rho^2} \right). \tag{13}$$

6   The VEV landscape

Therefore, according to Eq. (12), the σ field VEV seen by the accelerating (comoving) observer is position dependent. This is not strange since the proper acceleration along the x direction is breaking the Minkowski translation (and rotation) invariance. Now let us assume a comoving frame acceleration a belonging to the interval $0 < a < a_c$. The squared VEV is a function on the coordinate x ranging from v^2 for $x = \infty$ to zero, which is reached at some negative x value given by:

$$x_c = -\frac{1}{2a} \log \frac{a_c^2}{a^2} < 0 \tag{1}$$

where the phase transition takes place. Notice that the locus $x = x_c$ is indeed a surface because of the two other spatial dimensions $x_\perp$ which are free since the VEV is $x_\perp$ (as well as t) independent.

By using the approximation in Eq. (8) one finds:

$$x_c \simeq -\frac{1}{2a}(\frac{a_c^2}{a^2} - 1) < 0. \tag{2}$$

In this case one has to consider also a second critical value $x = x_c'$ where the squared VEV equals the asymptotic value v^2:

$$x_c' = \frac{1}{2a} > 0. \tag{3}$$

Obviously this approximation is useful only in the region $x_c < x < x_c'$ at most.

Now it is possible to write to $\bar{\sigma}^2$ in terms of the Minkowski coordinates X and T:

$$\bar{\sigma}^2 = v^2 \left(1 - \frac{1}{a_c^2 \rho^2} \right) = v^2 \left(1 - \frac{1}{a_c^2 (X^2 - T^2)} \right). \tag{4}$$

It is very interesting to realize that this function does not depend on the acceleration a but only on v and the critical acceleration a_c (which depends only on v and on N). In other words the VEV landscape depends only on the

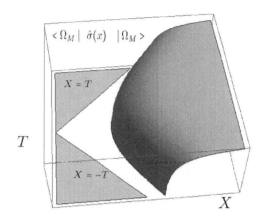

Figure 14: VEV of $\hat{\sigma}$ in Minkowski vacuum for different points of the spacetime, as seen by the accelerating observer.

parameters defining the LSM, but not on the acceleration of the comoving observer. In Fig. 14 we can find a plot of the VEV on the Minkowski as seen by the accelerating observer.

On the other hand we also have for the Minkowski quantum field $\hat{\sigma}_M(X)$:

$$< \Omega_M \mid (\hat{\sigma}_M(X))^2 \mid \Omega_M >= v^2. \tag{5}$$

At this point one may wonder; as σ is an scalar and, at the classical level, one should have:

$$\sigma(x) = \sigma_M(X) \tag{6}$$

on $\mathcal{R}$. Is this not in contradiction with Eq.(4) and Eq.(5)? The answer clearly is not, since:

$$< \Omega_M \mid (\hat{\sigma}_M(X))^2 \mid \Omega_M > \neq < \Omega_M \mid (\hat{\sigma}(x))^2 \mid \Omega_M > . \tag{7}$$

The reason is that $\hat{\sigma}_M(X)$ is an operator defined on the Minkowski Hilbert space $\mathcal{H}_M = \mathcal{H}_L \otimes \mathcal{H}_R$ where $\mathcal{H}_L$ and $\mathcal{H}_R$ are the Hilbert spaces corresponding to the regions $\mathcal{L}$ and $\mathcal{R}$ respectively. However $\hat{\sigma}(x) = \hat{\sigma}_R(x)$ is an operator defined only on $\mathcal{H}_R$, and it must be understood as $1 \otimes \hat{\sigma}_R(x)$ when acting on $\mid \Omega_M >$. An event belonging to the region $\mathcal{P}$ can affect events both in $\mathcal{L}$ and $\mathcal{R}$. Thus if $X_L \in \mathcal{L}$ and $X_R \in \mathcal{R}$, $< \Omega_M \mid \hat{\sigma}_M(X_L)\hat{\sigma}_M(X_R) \mid \Omega_M >$ does not necessarily vanish. This shows that that σ_M is not the tensorial product of σ_L and σ_R i.e.

$$\hat{\sigma}_M(X) \neq \theta(-X)\hat{\sigma}_L(x) \otimes \theta(X)\hat{\sigma}_R(x) \tag{8}$$

and then Eq.(4) and Eq.(5) are not incompatible at all.

7 Conclusions

The Unruh effect is an unavoidable consequence of QFT for accelerating observer. It applies to interacting theories and to any kind of fields (scalar, fermionic, gauge, etc). It can give rise to collective non-trivial phenomena such as phase transitions. In particular, in this work we have shown that a continuous spontaneously broken symmetry is restored for an accelerating observer. For her the VEV of the field depends on the position and it vanishes beyond a surface in the horizon direction. We conclude that all these facts are a solid evidence in favor of the ontic character of the Unruh effect.

Acknowledgments

The author thanks Vesselin Petkov and all the participants at the Fourth International Conference on the Nature and Ontology of Spacetime, Varna, Bulgaria (2016), and Luis Álvarez-Gaumé for comments concerning [15]. Work supported by Spanish grants MINECO:FPA2014-53375-C2-1-P and FPA2016-75654-C2-1-P.

References

[1] S.W. Hawking, Nature 248 (1974) 30; Comm. Math. Phys. 43 (1975) 199; Phys. Rev. D14 (1976) 2460.

[2] W.G. Unruh, Phys. Rev. D14 (1976) 870.

[3] L. C. B. Crispino, A. Higuchi and G. E. A. Matsas, Rev. Mod. Phys. **80**, 787 (2008).

[4] T. D. Lee, Nucl. Phys. B **264**, 437 (1986). R. Friedberg, T. D. Lee and Y. Pang, Nucl. Phys. B **276**, 549 (1986).

[5] S. Fulling, Phys. Rev. D7 (1973) 2850 .

[6] N.D. Birrell and P.C.W. Davies, Quantum fields in curved space (Cambridge University Press, 1982).

[7] L. Parker, Phys. Rev. D12 (1976) 1519.

[8] D.G. Boulware, Phys. Rev. Dll (1975) 1404; Phys. Rev. D13 (1976) 2169.

[9] R. Kubo, J. Phys. Soc. Jap. **12**, 570 (1957).

[10] P. C. Martin and J. S. Schwinger, Phys. Rev. **115**, 1342 (1959).

[11] J. Earman, Stud. Hist. Phil. Sci. B **42**, 81 (2011).

[12] C. Barcelo, S. Liberati and M. Visser, Living Rev. Rel. **8**, 12 (2005) [Living Rev. Rel. **14**, 3 (2011)]

[13] W. G. Unruh, Phys. Rev. Lett. **46**, 1351 (1981).

[14] L. J. Garay, J. R. Anglin, J. I. Cirac and P. Zoller, Phys. Rev. Lett. **85**, 4643 (2000).

[15] J. S. Bell and J. M. Leinaas, Nucl. Phys. B **284**, 488 (1987).

[16] A. Einstein, B. Podolsky and N. Rosen, Phys. Rev. 47 (1935) 777.

[17] T. Ohsaku, Phys. Lett. B **599**, 102 (2004).

[18] D. Ebert and V. C. Zhukovsky, Phys. Lett. B **645** (2007) 267.

[19] P. Castorina and M. Finocchiaro, J. Mod. Phys. **3** (2012) 1703.

10 SPACETIME IN EVERETT'S INTERPRETATION OF QUANTUM MECHANICS

LOUIS MARCHILDON

Abstract   Everett's interpretation of quantum mechanics was proposed to avoid problems inherent in the prevailing interpretational frame. It assumes that quantum mechanics can be applied to any system and that the state vector always evolves unitarily. It then claims that whenever an observable is measured, all possible results of the measurement exist. This assertion of multiplicity has been understood in many ways by proponents of Everett's theory. Here we shall illustrate how different views on multiplicity carry onto different views on spacetime.

1 Introduction

Everett's interpretation of quantum mechanics [1, 2] was proposed in a context where challenges to quantum interpretational orthodoxy were not well received [3]. At the time, the prevailing view drew both from the Copenhagen distinction between the quantum and the classical and from the Dirac–von Neumann collapse of the state vector.

Everett's framework attempts to do away with the collapse of the state vector, and to correct for a number of unsatisfactory aspects of the Copenhagen interpretation. The latter requires, in particular, a classical world logically prior to the quantum world, as well as observers outside quantum systems under investigation. In contradistinction to this, Everett's framework incorporates the following characteristics:

1. The state vector always evolves unitarily.

2. The 'observer' is included in the quantum description.

3. When a quantum measurement is performed, all results of the measurement occur.

A. S. Stefanov, M. Giovanelli (Eds), *General Relativity 1916 - 2016. Selected peer-reviewed papers presented at the Fourth International Conference on the Nature and Ontology of Spacetime, dedicated to the 100th anniversary of the publication of General Relativity, 30 May - 2 June 2016, Golden Sands, Varna, Bulgaria* (Minkowski Institute Press, Montreal 2017). ISBN 978-1-927763-46-9 (softcover), ISBN 978-1-927763-47-6 (ebook).

4. 'Collapse' happens relative to the observer only.

5. Quantum mechanics applies to the whole universe.

This general framework was, to some extent, articulated in Everett's published work which, however, left a number of questions only partly answered. Over the years, three problems have developed and received considerable attention from proponents of (or opponents to) Everett's approach. They have to do with probability, the preferred basis and multiplicity.

2   Three problems in Everett's approach

2.1   Probability

The probability problem in Everett's approach is twofold. The first horn of the problem concerns determinism, the second one has to do with numerical values.

As indicated above, according to Everett the state vector always evolves unitarily, in accordance with the Schrödinger equation. So the question is, How can probability arise if evolution is completely deterministic? The answer given by Everettians is that probability is not objective. Rather, it represents an observer's subjective uncertainty as to his post-measurement situation.

In quantum mechanics, numerical values of probability are given by the Born rule. Everett claimed that the Born rule does not need to be separately postulated, but can be derived naturally from the quantum formalism. The derivation turned out to be not so straightforward. Later investigators attempted to show that a rational agent who believes he or she lives in an Everettian universe will make decisions as if the square amplitude measure gave chances for outcomes. This has given rise to much debate [4], and the issue is not settled.

2.2   The preferred basis

Consider a two-state quantum system and an observable A with orthonormal eigenvectors $|a_1\rangle$ and $|a_2\rangle$. A measurement of A will involve an apparatus in an initial state $|\alpha_0\rangle$ which, upon interaction with the quantum system, will evolve to $|\alpha_1\rangle$ or $|\alpha_2\rangle$, respectively, if the system is prepared in $|a_1\rangle$ or $|a_2\rangle$.

If the quantum system's initial state is a superposition of $|a_1\rangle$ and $|a_2\rangle$, the system and apparatus will evolve as

$$(c_1|a_1\rangle + c_2|a_2\rangle)|\alpha_0\rangle \rightarrow c_1|a_1\rangle|\alpha_1\rangle + c_2|a_2\rangle|\alpha_2\rangle. \qquad (1)$$

But the final state can also be written as

$$c_1'|a_1'\rangle|\alpha_1'\rangle + c_2'|a_2'\rangle|\alpha_2'\rangle, \qquad (2)$$

with $|\alpha_1'\rangle$ and $|\alpha_2'\rangle$ orthogonal linear combinations of $|\alpha_1\rangle$ and $|\alpha_2\rangle$ (and $|a_1'\rangle$ and $|a_2'\rangle$ linear combinations of $|a_1\rangle$ and $|a_2\rangle$).

Now the question is, Why do we always observe well-defined macroscopic states $|\alpha_1\rangle$ and $|\alpha_2\rangle$, instead of ill-defined macroscopic states $|\alpha_1'\rangle$ and $|\alpha_2'\rangle$? The most popular answer is that the property of decoherence [5], which is independent of any interpretational frame, favors well-defined macroscopic states.

2.3 Multiplicity

In Everett's framework, all results of a measurement occur. Here the question is, What does this statement mean from an ontological point of view? Everett was not entirely clear on how to answer this question, at least in his published work. His followers have developed three distinct types of answer:

1. Many worlds: The whole universe splits in different copies.

2. Many minds: Although apparatus and brain states superpose, consciousness is well-defined.

3. Reality is identified with patterns in the universal wave function.

3 Interpreting quantum mechanics

Everett's approach provides an interpretation of quantum mechanics. Of course, there are many others [6]. But what does it mean to interpret quantum mechanics? According to the semantic view of theories [7, 8] it consists in answering the question, How can the world be for quantum mechanics to be true?

From this point of view, the problem of multiplicity is more pressing than the other two [9]. Indeed simply postulating the Born rule, as is usually done, will make probability well defined. And the preferred basis problem can also be solved by a specification, adequately guided by results from decoherence theory.

The extent of the ontological problem of multiplicity, gathered from the substantial literature on Everett's approach, has been analysed in [10]. The purpose of this paper is to see how the problem of multiplicity carries onto the ontological problem of spacetime.

4 Many worlds

As we pointed out above, Everett's views on multiplicity are not fully clarified in his published work. Probably the most straightforward way to understand multiplicity is to associate it with a literal split. That idea was popularized by DeWitt [11]:

> This universe is constantly splitting into a stupendous number
> of branches, all resulting from the measurementlike interactions
> between its myriads of components. Moreover, every quantum
> transition taking place on every star, in every galaxy, in every
> remote corner of the universe is splitting our local world on earth
> into myriads of copies of itself.

This quote suggests that a split occurs everytime an interaction produces entanglement, whether or not macroscopic objects are involved. Everett, however, introduced multiplicity only in contexts where something like a macroscopic apparatus performs a measurement. Considerations will henceforth be restricted to such situations.

Healey [12] was the first to formalize the consequences of splitting on the nature of space. He first considered the possibility that systems split into n copies, in usual ordinary space. But then, so he argues, mass-energy would be multiplied by n and we would presumably be aware of the overcrowding of space. Accordingly, the split can become acceptable in two different ways:

1. The physical systems do not split, only their states do.

2. Not only systems, but space itself splits. The resulting systems may be viewed as living in a higher-dimensional manifold.

The first way anticipates multiplicity viewed as decoherent sectors of the wave function, which we will examine later. The second way is perhaps the easiest one to visualize. It involves multiple copies of spacetime, all of them multiplying further upon each measurement interaction. It does, however, raise a number of questions usually associated with state vector collapse: When, in the measurement process, does the split occur? Does the split occur on an equal-time hypersurface, or on a light cone (so as to be more consonant with special relativity)? What are the precise conditions that define a measurement?

Related to splitting is the intriguing question of recombination. If evolution is unitary, measurements can in principle be undone. Just like worlds, multiple copies of spacetime should then recombine. This, however, can be avoided by means of bifurcation, an alternative to splitting introduced by Deutsch [13]. The idea is that there are infinitely many worlds (or spaces) at any time. Their number neither increases nor decreases. In measurement contexts the set of all spaces is partitioned in as many subsets as there are possible measurement results. In this case the multiplicity of the whole universe does not change, nor does the complete spacetime arena.

5 Many minds

Albert and Loewer [14] gave the first full-fledged formulation of the idea that the split involves the mind rather than the world. In the many-minds view, every observer has associated with it an infinite set of minds. Minds are associated with brain states but are not subject to superposition.

Many-minds approaches usually involve a single spacetime, in which different experiences coexist. But according to Lockwood [15],

> the fact of a physical system's being in a superposition, with respect to some set of [consciousness] basis vectors, is to be understood as the system's having a *dimension* in addition to those of time and space.

Just as one's mind is usually thought as being wholly present at different times when it has different experiences, it can also be thought, according to Lockwood, as wholly present in each different experiences it can have at a single time. These experiences can be viewed as lying on an axis orthogonal to the time axis. Spacetime is therefore enlarged, but by a dimensionality that is neither spatial nor temporal. Lockwood does not speculate on how the spacetime metric can connect to the additional dimension.

6   Patterns in the wave function

Decoherence theory is an important building block of an approach to Everett different from many worlds and many minds. It is mainly connected with the names of Gell-Mann and Hartle [16], Saunders [17] and Wallace [18, 19] (although Gell-Mann and Hartle favor one world in the end).

Wallace [18] identifies real structures with stable patterns in the universal quantum state:

> My claim is instead that the emergence of a classical world from quantum mechanics is to be understood in terms of the emergence from the theory of certain sorts of structures and patterns, and that this means that we have no need (as well as no hope!) of the precision which Kent and others here demand.

At the beginning of an experiment, there is one apparatus pattern in the universal quantum state. This, according to Wallace, means that there is one apparatus. At the end of the experiment, the universal quantum state, represented on the right-hand side of (1), contains two distinct apparatus patterns. Therefore there are now two apparatus. It doesn't make sense, according to Wallace, to ask questions about the apparatus in the very short decoherence timescale leading from one pattern to two patterns. Nor does the existence of two distinct patterns cause problems for, as one can see in Schrödinger's cat imagery:

> If A and B are to be 'live cat' and 'dead cat' then [the relevant microscopic properties] P and Q will be described by statements about the state vector which (expressed in a position basis) will concern the wave-function's amplitude in vastly separated regions R_P and R_Q of configuration space, and there will be no contradiction between these statements.

So the live cat and the dead cat occupy different regions in configuration space. But how do they project in three-dimensional space? In classical theory, two different cats not only occupy different regions in configuration space, but also different regions in three-space. That is, their projections from configuration space to three-space do not overlap.

This, however, is not so with Wallace's patterns. Projected in three-space, the two cats may literally overlap. How can one understand this?

One possible answer is to assume that the live cat and the dead cat don't project from configuration space to the same three-space [20]. This means, for instance, that there is an added parameter, or another dimension, introduced to distinguish different three-spaces from each other.

This, however, is not the answer that Wallace prefers. According to him, there is only one three-space into which both patterns project. This implies that the live cat and the dead cat are, so to speak, ghostlike to each other. In the words of Allori *et al.* [21], "[t]he two cats are [...] reciprocally transparent."

Do we have to choose between one three-space and many three-spaces? Investigating macroscopic ontology in Everettian quantum mechanics, Wilson [22] suggests that we may not:

> [T]he 'spacetime' of the quantum mechanics and quantum field theory formalism, in terms of which branches are defined, is not the same as the 'spacetimes' of macroscopic worlds. The former 'spacetime' is a single entity common to multiple branches, while each of the latter 'spacetimes' is tied to a particular macroscopic course of events.

But Wilson also draws an analogy between bifurcation and consistent histories, each history being located in a distinct spacetime.

7   Discussion

We have shown elsewhere [10] that there is a wide spectrum of opinions, among adherents to Everett's approach, on the nature of Everettian multiplicity. This, as we have just seen, translates into different views of space or spacetime, which fall into two broad categories:

1. Space (or spacetime) genuinely splits upon measurement of a quantum observable.

2. Space (and spacetime) don't split.

To some extent, this alternative occurs in all three main approaches to multiplicity, namely, many worlds, many minds and decoherent sectors of the wave function.

If space genuinely splits, the problems traditionally connected with state vector collapse seem to be carried along the way. This does not mean that

Everett's approach has to be rejected, but it invites Everettians to make their views substantially sharper.

If space doesn't split, we are confronted with various copies of macroscopic systems occupying the same spatial arena. The problem is, How can these systems not interact? In other words, How is the macroscopic multiplicity reconciled with the quantum field theory of interacting constituents? Wilson acknowledges the problem when he introduces two different notions of spacetime. The solution, however, remains to be implemented.

8 Conclusion

Interpreting quantum mechanics consists in answering the question, How can the world be for quantum mechanics to be true? Everett's approach is one possible answer to the question. Every answer currently considered has problems of its own. We conclude by pointing out three problems that are perhaps more specific to Everett's theory.

The first one is that it is not sharply defined. This has been documented here with respect to spacetime, and in [10] with respect to multiplicity. This contrasts with the de Broglie–Bohm approach which, at least in the nonrelativistic case, is rather well defined. Other interpretations do carry some measure of indefiniteness, for instance the Copenhagen distinction between the quantum and the classical. But the diversity of views among Everett's adherents is particularly striking.

The second one can be considered either a problem or a strength, depending on one's point of view. It consists in the fact that Everett's approach is highly dependent on the exact validity of quantum mechanics. Everett's many worlds, so it seems, won't survive the smallest nonlinear term to the Schrödinger equation. By contrast, again, the de Broglie–Bohm approach is highly adaptable to changes in the formalism of quantum mechanics [23].

The third problem has to do with Everett's extraordinary ontology. True, science has taught us that common sense is not always the best of guides. Yet, as in experimental investigations, "Extraordinary claims require extraordinary evidence." Many critics of Everett's approach believe that such evidence has not convincingly been put forward by Everett's supporters.

Acknowledgements

I am grateful to the Natural Sciences and Engineering Research Council of Canada for financial support.

References

[1] H. Everett III, " 'Relative state' formulation of quantum mechanics," *Reviews of Modern Physics* **29** (1957) 454–462.

[2] H. Everett III, "The theory of the universal wave function," reprinted in B. S. DeWitt and N. Graham (Eds.), *The many-worlds interpretation of quantum mechanics* (Princeton University Press, 1973), pp. 3–140.

[3] O. Freire Jr., *The quantum dissidents: Rebuilding the foundations of quantum mechanics (1950–1990)* (Springer, 2015).

[4] S. Saunders, J. Barrett, A. Kent and D. Wallace (Eds.), *Many worlds? Everett, quantum theory, and reality* (Oxford University Press, 2010).

[5] M. Schlosshauer, "Decoherence, the measurement problem, and interpretations of quantum mechanics," *Reviews of Modern Physics* **76** (2004) 1267–1305.

[6] D. Greenberger, K. Hentschel and F. Weinert (Eds.), *Compendium of quantum physics* (Springer, 2009).

[7] B. C. van Fraassen, *Quantum mechanics: An empiricist view* (Clarendon Press, 1991).

[8] L. Marchildon, "Why should we interpret quantum mechanics?" *Foundations of Physics* **34** (2004) 1453–1466.

[9] L. Marchildon, "Can Everett be interpreted without extravaganza?" *Foundations of Physics* **41** (2011) 357–362.

[10] L. Marchildon, "Multiplicity in Everett's interpretation of quantum mechanics," *Studies in History and Philosophy of Modern Physics* **52** (2015) 274–284.

[11] B. S. DeWitt, "Quantum mechanics and reality," *Physics Today* **23**(9) (1970) 30–35.

[12] R. A. Healey, "How many worlds?" *Noûs* **18** (1984) 591–616.

[13] D. Deutsch, "Quantum theory as a universal physical theory," *International Journal of Theoretical Physics* **24** (1985) 1–41.

[14] D. Albert and B. Loewer, "Interpreting the many worlds interpretation," *Synthese* **77** (1988) 195–213.

[15] M. Lockwood, *Mind, brain and the quantum: The compound 'I'* (Blackwell, 1989).

[16] M. Gell-Mann and J. B. Hartle, "Quantum mechanics in the light of quantum cosmology," in W. H. Zurek (Ed.), *Complexity, entropy, and the physics of information* (Addison-Wesley, 1990), pp. 425–458.

[17] S. Saunders, "Time, quantum mechanics, and decoherence," *Synthese* **102** (1995) 235–266.

[18] D. Wallace, "Everett and structure," *Studies in History and Philosophy of Modern Physics* **34** (2003) 87–105.

[19] D. Wallace, *The emergent multiverse. Quantum theory according to the Everett interpretation* (Oxford University Press, 2012).

[20] G. Bacciagaluppi, "Remarks on space-time and locality in Everett's interpretation," in T. Placek and J. Butterfield (Eds.), *Non-locality and modality* (Kluwer, 2002), pp. 105–122.

[21] V. Allori, S. Goldstein, R. Tumulka and N. Zanghi, "Many worlds and Schrödinger's first quantum theory," *British Journal for the Philosophy of Science* **62** (2011) 1–27.

[22] A. Wilson, "Macroscopic ontology in Everettian quantum mechanics," *Philosophical Quarterly* **61** (2011) 363–382.

[23] A. Valentini, "De Broglie-Bohm pilot-wave theory: Many worlds in denial?" in [4], pp. 476–509.

Part III

FORMAL APPROACHES TO SPACETIME

A. S. Stefanov, M. Giovanelli (Eds), *General Relativity 1916 - 2016. Selected peer-reviewed papers presented at the Fourth International Conference on the Nature and Ontology of Spacetime, dedicated to the 100th anniversary of the publication of General Relativity, 30 May - 2 June 2016, Golden Sands, Varna, Bulgaria* (Minkowski Institute Press, Montreal 2017). ISBN 978-1-927763-46-9 (softcover), ISBN 978-1-927763-47-6 (ebook).

11 An Abstract Ontology of Spacetime

Thomas Benda

Abstract   An abstract ontology of spacetime is proposed. It is the model of a conservative extension of Zermelo's Z, which provides an axiomatic construction of (general) relativistic spacetime performed in previous work (Benda 2008, 2015, 2017), using a conceptual base that includes no more than worldlines and a ternary intersection relation. The ontology is an extended set-theoretical hierarchy, in which mathematical entities arise out of the empty set and physical entities arise out of urelemente. The urelemente are interpreted as structureless worldlines, each representing continuous flow of time. Space is secondary to time by having spacetime points as sets of worldlines and slices of space as sets of spacetime points. The proposed ontology rests on a platonist view of the mathematical world of sets. By admitting urelemente which represent continuity, a platonist combined mathematico-spatiotemporal world is created.

1   Introduction

In a common picture, suggested by our daily experience, spacetime is the stage of the physical world, the setting in which physical events unfold. The actors on the stage are material objects, that is, matter and fields. There may be no stage without actors, as the followers of Leibniz claim, and only actors may act, that is, cause direct experience to the spectator. But even then, properties may be well attributed to the stage itself, independent of who or what fills it. In the case of spacetime, such properties include dimensionality and causality.

Spacetime is readily described precisely in a mathematical language. Accordingly, as soon as quantitative physical laws were formulated, spatial and temporal coordinates appeared in them. But it was not before the Special Theory of Relativity that spacetime itself became the main subject of a physical theory. Even though the benefits of axiomatizing physical theories occasionally have been doubted, over the years, axiomatizations of both

A. S. Stefanov, M. Giovanelli (Eds), *General Relativity 1916 - 2016. Selected peer-reviewed papers presented at the Fourth International Conference on the Nature and Ontology of Spacetime, dedicated to the 100th anniversary of the publication of General Relativity, 30 May - 2 June 2016, Golden Sands, Varna, Bulgaria* (Minkowski Institute Press, Montreal 2017). ISBN 978-1-927763-46-9 (softcover), ISBN 978-1-927763-47-6 (ebook).

the Special Theory of Relativity and the General Theory of Relativity were undertaken by, among others, Reichenbach (1924), Robb (1936), Ax (1978), Mundy (1986), Ehlers, Pirani and Schild (1972) as well as Hawking et al. (1976) and Malament (1977a, b). In recent years, interest in precise logical formulations of spacetime theories has been rekindled by a group around István Németi, who set out to formulate axiomatic theories of spacetime equipped with idealized particles and observers in a spirit of algebraic logic. Axiomatizations were provided for variants of the Special Theory of Relativity (H. Andréka, J.X. Madarász and I. Németi 2004, 2006, 2007) and, in a later step that incorporated accelerated observers, the General Theory of Relativity (J.X. Madarász, I. Németi and Cs. Tőke 2004); H. Andréka, J.X. Madarász, I. Németi and G. Székely 2008, 2012). The theories are motivated by Einstein's Relativity and Equivalence Principles and are formulated in various versions, whose underlying assumptions, mutual dependence and implications are compared. A relatively sparse first order language is employed and the sole emphasis lies on mathematical structure, without ontological presuppositions, neither mathematical nor physical, so that entities like particles, photons and observers are characterized only by their spatiotemporal coordinates and, if applicable, relative values of their masses.

As a complement to the approach by Németi's group, one may place ontology at the center of interest, regarding both mathematical entities and spacetime itself. For philosophers, an ontological interest is warranted. The bare and empty character of spacetime suggests a more intimate connection of spacetime to mathematical entities than to physical objects, which gives us a motive to look at both, at least in loose conjunction.

A wide, not strictly one-dimensional spectrum of views is available between declaring both mathematical entities and spacetime to be mere mental constructs and taking that as a reason to drop the subject from ontological discourse and, at the other end, platonism with respect to abstract entities. We will not discuss the merits of the various views and the reservations against them, that would go beyond the present limited scope. We will rather stay close to the platonistic end of the spectrum and show its feasibiliy. To carry that through, we do not need to do more than laying out an ontological foundation of spacetime which promises to be as simple as possible, assuming as little as possible and seeing how far it carries us. More specifically, we will construct, in precise terms, relativistic spacetime as a structure that is ontologically and conceptually parsimonious. Subsequently, we will establish a connection to the structure of the mathematical entities, that is, the set-theoretical hierarchy. Finally, we will present a primitive criterion of being not purely mathematical, which will apply to time: continuity in an original and intuitive sense, which as been lost in modern scientific discourse. All entities involved will be abstract in the sense that no ad hoc or instrumentalist entities will be employed and that no reference to the concrete world will be required. Most of the work is technical, using first order logic. Presently, however, for brevity and readability, we will make do with vernacular descriptions. Those tasks will be dealt with in turn in the following sections.

2   A parsimonious structure

Spacetime, in a mathematical representation, is commonly introduced as a pair of a manifold and a metric, which is usually denoted by $\langle M, g \rangle$. To be in accordance with physical experience, M is four-dimensional and g is Lorentzian. Thus, on the one hand, one temporal and three spatial dimensions are accommodated and, on the other hand, causality is respected. Yet both M and g are fairly complex and so one may seek to reduce them to simpler entities. Preferably, the sought primitive entities are closely associated with properties that we readily deduct from our experience and attribute to spacetime itself. As such a property, causality stands out. In the common mathematical representation, causality emerges at a comparatively high level: two spacetime points are causally connected if and only if their distance according to g has a positive value. Causality is preserved under diffeomorphism and, as its name suggests, is the determinant of the possibility of causal connections between material objects. Theefore, it seems appropriate to place causality at the foundation of any construction of spacetime and have it incorporated in the basic entities thereof.

It looks natural to make spacetime points the primitives of spacetime. In such a set-up of spacetime, causality, if understood to be basic, would be a binary relation of spacetime points. We will define such a relation below, read "p_1 causally precedes p_2", in symbols, $p_1 \prec p_2$. However, we aim to be more radical. Not spacetime points, but entities that represent causal flow and are called "worldlines" will be the primitive entities of spacetime. Our terminology here and in another instance further below differs from standard usage and is summarized in an appendix. Worldlines are readily visualized as connecting spacetime points, precisely those spacetime points which are causally connected. However, such a visualization presupposes entities and structure we do not have yet. With worldlines as primitives, at this initial stage, there is no geometric space in which worldlines are placed, in particular, there are no distances, angles or intersecting geometrical objects, not even spacetime points. Only later, our visualization will be justified.

As formal objects, worldlines are structureless. They contain no elements. Yet they are capable of undergoing a relation we call "intersection". We say of two worldlines v and w which are thereby related "w intersects v". In order to bring in our pre-formal intuition of worldlines as causal flow, we wish to have w intersect v with an additional parameter which represents a notion of "earler" or "later". So we extend the intersection relation to three places, saying "w intersects v at x", in symbols, $v \,\big\slash_x\, w$, where x is an element of a totally ordered set. Later in our construction, objects of differential geometry have to be defined in a standard way. The—totally ordered—set of real numbers will allow that in the simplest and coarsest way.

With that in mind, a formal theory, denoted by ST, is set up (Benda 2008, 2015, 2017). ST is a conservative extension of Zermelo's set theory Z. Its primitive vocabulary, in addition to that of Z, contains no more than an additional ternary predicate constant, $\big\slash$. More details will be given in the following subsection.

The language of ST is expressive enough to establish relativistic space-time. By definition and with the help of the axioms of ST, the known entities of spacetime are roughly constructed as follows.

Given a worldline v and a real number x, a spacetime point is defined as the set of the worldlines intersecting v at x united with v. By the axioms of ST, every worldline w that intersects v at x is intersected by v at some real number y. Furthermore, every worldline that intersects v at x intersects w at y. So we can speak of mutual intersection of worldlines. Every pair $\langle v, x \rangle$ of a worldline v and a real number x corresponds to precisely one spacetime point p. Hence all spacetime points containing v form a totally ordered set, which we call a "worldline curve". A little sloppily, but suggestively, we speak of a "real number x on a worldline v", meaning the corresponding unique space-time point that is an element of the worldline curve v^* corresponding to v. Spacetime points on a common worldline curve are causally connected to each other and are totally ordered. For every pair of spacetime points p_1, p_2, we have that either p_1 causally precedes p_2, in symbols, $p_1 \prec p_2$; or p_1 causally succeeds p_2, its converse; or p_1 and p_2 are causally not connected. We define the causal future and past of any given spacetime point p as the set of space-time points that succeed and precede p, respectively. They are analogues of the known lightcones of relativistic spacetime, but without any geometric structure yet. Intersections (here and only here, in the set-theoretical sense) of futures and pasts have the properties of open sets, so a topology is readily defined, which resembles the Alexandroff topology, but again without any geometrical structure yet. With a topology in place, we are able to speak of worldline curves running through neighborhoods, partiularly, through disjoint neighborhoods in sequence. Furthermore, we are able to define any property which is functionally dependent on spacetime points as local for said property obtaining in some neighborhood.

As a next step, by the axioms of ST, coordinates of spacetime points can be defined. For any given spacetime point p, there are two worldlines v and w, the associated worldline curves v^* and w^* of each of which run through the past of p, then through spacetime points causally not connected with p, and then through the future of p. Thereby, both on v^* and w^*, we have intervals of real numbers that do not correspond to spacetime points causally connected with p. Both intervals have suprema and infima. They are unique in a neighborhood of p and vary smoothly within it, so can be employed to define coordinates of p with respect to v and w. By the axioms of ST, we obtain an atlas of smoothly overlapping coordinate systems, which we call "standard spacetime atlas".

Thus we have established the spacetime manifold. It is four-dimensional since the coordinate systems of the standard spacetime atlas are defined with respect to two worldlines each. Using only one worldline for the definition of coordinate systems would result in a two-dimensional spacetime manifold.

Further on, we are able to define common entities of differential geometry (O'Neill 1983) solely in the language of ST: tangent vectors, tangent vector fields, smooth manifold transformations, tangent vector transforma-

190

tions; from there, curves in spacetime, velocity vectors of curves; and finally, connections, accelerations of curves (as dependent on connections and parametrizations of curves) and geodesics. Worldline curves are defined as above and parametrized in a natural way by real numbers on them, that is, with v being a worldline, v^* being is associated worldline curve, p being a spacetime point, $v \in p$ and $\langle v, x \rangle$ corresponding to p, p is the value of v^* at x. By the axioms of ST, worldline curves are curves in spacetime with respect to the standard spacetime atlas. Furthermore, locally, there is a partition of worldlines into sets such that the velocity vectors of the corresponding worldline curves of each set form a tangent vector field. That allows to consider how any given connection produces accelerations of all worldline curves. The axioms of ST put restrictions on how parametrizations of different worldline curves are related to each other. The restrictions are such that there is a single connection which renders all worldline curves acceleration-free, that is, geodesics. It is called "geodesic connection".

Finally, the geodesic connection is used to construct a metric as a 0-2 tensor derivation in a known procedure (O'Neill 1983). Technically, the metric is the 0-2 tensor derivation constructed out of any tangent vector field t (which is a 0-0 tensor derivation) and the geodesic connection (from which, given t, a 1-0 tensor derivation is obtained) whose kernel vanishes for all t. The metric, in particular, its signature, is determined up to a factor that is constant in spacetime. A signature of the obtained metric is not obtained by the axioms of ST, but can be motivated as follows. We first observe that, for any spacetime point p, the four-dimensional manifold of coordinates of all worldline curve velocity vectors at p is convex. Then we consider diffeomorphisms on spacetime and among them a set d_1 whose elements preserve causality as well as a set d_2 whose elements preserve the sign of the squared length of any tangent vector under the found metric. Both d_1 and d_2 can well be taken to preserve physics. Stipulating $d_1 = d_2$ results in the found metric to have Lorentzian signature.

Above construction of relativistic spacetime is lengthy, but rests conceptually, besides set theory, on no more than worldlines and intersection thereof by other worldlines plus several conditions on intersection and linearly ordered places thereof, provided in the shape of fifteen axioms. The notion of spacetime manifold with metric, commonly taken to be basic, has thereby been reduced to something simpler, which, nevertheless, is a destillate of an important feature of spacetime, causality.

3   The extended set-theoretical hierarchy

Putting an important physical theory like the Theory of General Relativity on an axiomatic foundation is motivated by a desire of precision and of the ability to apply mathematical tools to all its ranges. Axiomatizations of physical theories usually consist of stipulations ("axioms") in mathematical language, on the one hand, and of establishing correspondences between mathematical and physical entities in natural language, on the other hand (see, e.g., Bunge 1967), which are often merely implied rather than explicitly stated. In such axiomatizations, clearly not every mathematical entity has a physical counterpart or, as we may say, a physical interpretation. For example, in an axiomatization of classical mechanics, triples of real numbers and their first and second derivatives are respectively interpreted as positions, velocities and accelerations, but higher derivatives remain without physical counterparts.

Assuming that there are physical objects, such as particles and fields, as we do in common interpretations of classical mechanics, field theory and quantum mechanics, we find that it is not physical objects which have mathematical counterparts, but, in a generalized sense, properties thereof. That is clear in classical physics. Besides, the idea of physical objects—particles or fields—has survived the advent of quantum mechanics. Monadic properties of physical objects are given in mathematical form, whereas physical objects themselves are poined to by indices, for example, assignments of dimensions in an Hilbert space. The discussion of the topic of physical objects and their properties in quantum mechanics is, of course, more complex than we can indicate in this brief brush (see, e.g., French and Krause 2006, Ch. 4). In particular, entangled states pose a challenge in that they exhibit properties of the whole, but not of constituents. Yet denying that there are no constituents just because we are unable to isolate monadic properties that may be assigned to them (Margenau 1944) or engineer a quantum collapse that affects only one of them seems to be premature. In the Hilbert space of an entangled state, constituents appear as subspaces, even though they have no separable monadic properties. Therefore, the repository of physical entities in physical discourse does not emerge from the axiomatization in question, but has to be put into it, motivated and justified by pre-axiomatic considerations, mainly results—better: interpretations—of experiments and earlier not yet axiomatized parts of the theory to be set up.

What makes objects physical rather than mathematical is, in a simple understanding, their concreteness, their being subject to immediate experience rather than their being abstracted therefrom in thought. That inescapably seems to imply physical objects being there in space and time. With all that, any axiomatization of a physical theory in a systematic order involves establishing a range of physical objects of interest, assigning them properties (again, in a generalized sense), mathematizing those properies, and stipulating, in mathematical language, formulas they satisfy.

Above simple remarks set the stage for how to axiomatize a theory of spacetime. What physical objects does spacetime—without anything in it—involve? It is tempting to say: none (Brown and Pooley 2006), since spacetime itself is not the subject of our experience. Indeed, physical laws concern bodies and fields rather than space and time and only the former contribute to the explanatory work we expect from our theories.

We may, however, take serious our way of speaking according to which physical events occur in spacetime rather than falling under a concept of spacetime (Kant 1998, A25/B40). We presently follow such an approach and give spacetime a non-mathematical quality, which we take to be a physical quality and which will be motivated further in the next section.

Proceeding on that path, we first note that, according to a common view, mathematical structures, tuples of mathematical objects, properties and relations, are not plainly there. There is no need to specify them beyond a capability of being isomorphically mapped into each other and to satisfy certain theorems, so that, in the case of numbers, "[i]t is in this sense [i.e., we are unable to say whether we talk about certain mathematical objects or their Gödel numbers] true to say, as mathematicians often do, that arithmetics is all there is to number" (Quine 1969, p. 45). Under such a view, mathematizing the properties of spacetime is straightforward and even hardly noticed as a step. Spacetime is then adequately described as just such a structure, viz., as a four-dimensional manifold with a certain metric. If, however, we employ a mathematical ontology, we will be reluctant to identify physical spacetime with a mathematical object.

Mathematical platonists, a minority among philosophers of mathematics, construct a mathematical ontology in the shape of the set-theoretical hierarchy. We briefly review its setting up, following Ulrich Blau (2008). Crucial, non-eliminable, not definable without circularity (Savellos, 1990), and yet understood by everyone, is the notion of identity. We call the capability of objects to be identical with themselves and distinct from others "individuality". Individual objects are collected in sets. By elimination of objects from any set, we obtain subsets thereof and are, if we continue to do so, finally left with nothing. We call any set which is individuated, contains only individuated elements and to which a cardinal extension can be given, "extensionally determined". With that, we state two judgments, which are a priori, unprovable and irrefutable, and seemingly are presupposed in much, possibly everything, of what we state concerning individuals.

(S0)   Every set is uniquely determined by its elements and is individuated provided all of its elements are individuated.

(S1)   The collection of subsets of every extensionally determined set is extensionally determined.

The nothing is objectified as the empty set. By (S0), it is unique. While looking innocent, (S0) may be reckoned to be too restrictive in that it makes the nothing absolute. Could there be more than one empty set, obtained by taking out elements from a given set? If so, if there were two empty sets, say, $\varnothing_1$ and $\varnothing_2$, we could form the set which contains precisely those, in common denotation, $\{\varnothing_1, \varnothing_2\}$, and again remove both of its elements, ending up with nothing, which cannot be $\varnothing_1$ but not $\varnothing_2$, or vice versa. Here we still assumed that (S0) holds well for sets with something in them. For any additional parameter that makes sets with fixed elements not unique would have to be an additional element or something beyond being a mere individual.

The absoluteness of the nothing provides a fixed anchor for the set-theoretical hierarchy defined below. Furthermore, by (S0) and (S1), power sets and unions of extensionally determined sets are extensionally determined. With that, we are able to construct the set-theoretical hierarchy by the following self-referential imperative.

(V0)   Consider the empty set.

(V1)   Consider the power set of the set considered in the last step and repeat (V1).

(V2)   Consider the union of all sets to be considered that way and repeat (V1) and (V2).

In this build-up, a limit number level is reached each time (V2) is employed.

The set-theoretical hierarchy is not extensionally determined, since, being complete, it has no cardinal extension. In particular, as is well known, it is not a set, but a proper class. Its construction is not descriptive, but proceeds by imperative towards an intended whole. We speak of the hierarchy as "intensionally determined" and denote it by V.

While we do not engage into arguing for mathematical platonism, we again note that no mathematical theory and, as far as one can see, no treatise of individuals does without presupposing the ability of sets to be formed as well as above judgments (S0) and (S1). Mathematical platonism may be rejected, but its rejecting does not simplify our foundation of mathematics or decrease the number of presuppositions we need therefor.

Spacetime, viewed as a physical object, is not part of the set-theoretical hierarchy V, nor are wordlines, the constituents of spacetime in our model which were introduced in the previous section. Nevertheless, worldlines are individuals and are thus capable of forming sets. Since, by (S0), there are sets whose elements are worldlines and elements of V, we do not have a hierarchy of sets based on worldlines which is separate from V. Instead, we construct a combined set-theoretical hierarchy by a self-referential imperative consisting of steps (U0), (U1) and (U2):

(U0)   Consider the set of worldlines.

(U1)   Consider the power set of the set considered in the last step and repeat (U1).

(U2)   Consider the union of all sets to be considered that way and repeat (U1) and (U2).

The thereby obtained combined set-theoretical hierarchy is denoted by U. The class V is a subclass of U. The elements of the set we start out with are, as is common, called "urelemente". Urelemente are, in U, featureless individuals, without elements of their own. They represent objects whose characteristics are formally inexpressible and thus fall outside the pure structure of distinctness and elementhood. Accordingly, sets in V, those which arise from the empty set alone are called "pure sets". The remaining sets in U are called "impure sets".

Interpreting the elements of V and $U \setminus V$ is now straightforward. The former are mathematical objects and the latter are physical objects. Of course, only tiny parts of V and $U \setminus V$ on low levels of the hierarchy will be of practical importance.

Integrating physical objects as urelemente into set theory is, of course, not a far-fetched idea. Usually, such an endeavour is hampered by the messy individuality of physical objects. It promises to work only for elementary particles or other primitive constituents of the physical world. Even then, such an approach, including the presupposition of individuality of the physical urelemente, would be heavily dependent on the adopted physical theory. Here, however, we are proposing urelemente whose individuality is baked in from the beginning and which, as mere constituents of spacetime, will survive theory change that does not affect spacetime itself.

We thus have embedded our formal theory ST in a general setting, taking the notion of individuality serious and exercising a spirit of rigor. The basic constituents of the intended model of ST, of spacetime, are incorporated into an extended set-theoretical hierarchy, $\mathbb{U}$. At the same time, we seek to provide an ontological foundation of spacetime. Its basic entities are worldlines. Formally, worldlines are individual, structureless objects. What each of them represents, flow of time, is an informal notion, which we will explore in the next section.

4 Continuous flow of time

Continuity as a notion is apparently understood by every child. The dictionaries tell us that continuity is "the state or quality of being continuous" (Webster 2003), which, in turn, signifies "being in immediate connection or relation" (Webster 2003) or "without a pause or interruption" (Cambridge Dictionary 2013). What is continuous proceeds as a flow, without taking leaps between various stages. Movies are series of still pictures, which are screened in a succession rapid enough to create an illusion of continuity in the viewer. Such an illusion would not make sense if its purpose were no more than feigning an even more rapid succession of still pictures. Stroboscopic lighting in a discoteque serves the opposite purpose, to break down the natural impression of continuous movement into a sequence of discrete, discontinuous images. We distinguish between flow and jerky succession, not between rapid and slow succession of discrete impressions. The concept of flow does not require a concept of its stages. Similarly, we understand movement without relying on snapshots taken in between. Zeno's arrow is not moving by virtue of snapshots taken at ever closer time intervals showing ever closer spatial positions thereof. Snapshots show nothing but an arrow at rest at various stages.

Calculus, as developed by Dedekind, has taught us to view continuity as sufficient closeness of individual stages to each other. Accordingly, we speak of the continuity of real numbers. Here continuity presupposes individuality, as it befits any concept of mathematics, whose objects, after all, are individuals. The success of mathematics and mathematical science easily leads us to believe that there is no more to continuity than that. Indeed, one may argue that the idea of continuity presented in the preceding paragraph is nothing but an idealization of an ever denser succession of individual states, serving as a simplification, given our limited cognitive abilities and range of memory. That idealization is unnecessary for mathematics and science. Yet even if so, the device of continuous flow—now and henceforth in the sense of the previous paragraph—appears to us as quite ingenious. In one stroke, it eliminates the myriad unnecessary details of all possible snapshots and leaves us with the essential, movement itself, without any need of individuality.

Closer inspection reveals that continuity cannot even be approximated by ever denser arrangements of stages. We look at an ancient example. Geometry since the times of Euclid used to be about points, lines, planes and solid bodies. Neither were points thought to constitute lines, nor lines to constitute planes. With Cartesian analytical geometry and the acceptance of infinity into mathematics, the attitude changed. It is now asked: "How many points are there on a line in euclidean space?" (Gödel 1964). And yet, the length of points, which have no spatial extension, may be multiplied by any, even infinite, number to result in nothing but zero. No infinite arithmetics has been brought forward that dispenses with zero as the neutral element of multiplication. Pluralities of points do not create lines. Every common person is able to draw a line without any idea about individual points constituting that line. Schoolchildren are taught accordingly without difficulty. The difficulty would be greater, were they told that infinitely many points constitute those easily and readily drawn lines. Placing a multitude of points next to each other, no matter how high its cardinality is, does not create a line. We draw the line by producing an appropriate multitude of spots and specks, beholding it and subsuming it under our existing concept of a continuous line, just as we do with any of our drawings. Continuity as we understand it may be an illusion, but the illusion would not be possible without the existence of a concept that is not reducible to arrangements of individual entities.

To summarize, arrangements of points, no matter how dense, are neither necessary nor sufficient to bring about the continuity of a line. They are not necessary because our concept of a line precedes or at least is independent from our concept of infinite multitudes. They are not sufficient because, according to arithmetics, infinitely many spatially extensionless points taken together still have no spatial extension and because continuity is not expressible in terms of individuals.

The present proposal is to take the idea of continuity as just discussed serious and put it at the bottom of our set-up of spacetime.

What is continuous—henceforth understood as presented above, not as in calculus—is time. Our intuition strongly supports time as passing and fleeting, without ever stopping. Were time a totally ordered series of time points, for example, represented by the real numbers, then the world would be frozen into a totally ordered arrangement of spatial worlds. They would, if close enough to each other, resemble each other to any desired degree. But nothing would ever move or change.

Of course, in our physical laws, time just is represented by the set of real numbers or some other totally ordered set. Physical causality applies and evolutionary development bears out as usual in the theoretically suggested totally ordered arrangement of still worlds. Why would there be a need to stipulate continuity of time? Without continuity, spatial points in time become spacetime points, their set becomes spacetime, and only higher-order entities, such as tangent vectors, can be separated into timelike and spacelike entities. That works mathematically to formulate our physical theorems, but is blind to a different character of time in contrast to spatial directions which is suggested by how we perceive the physical world. Whatever the mechanism of our perception is, a topic that goes well beyond the present scope, space is presented to us directly as the stage of an arrangement of physical objects that partly persists over time and partly undergoes change therein. Change is predicated on flow of time. So time is presented to us indirectly, by way of movement on the spatial stage, through recording, memory and anticipation, for which participation of mental states is instrumental.

We do not know whether perceived changes and transitions of physical objects which are located in space have an entirely adequate mathematical description. Saying that seems to be at odds with the received view about change of spatial position. Again, don't we have calculus? Certainly, for our practice and given our best physical theories, calculus provides all we need. Still, calculus is no more than a tool that fairly accurately puts on a quantitative base what we have in mind when we think of movement: change of location. Without having a concept of transitional change, fading out of the previous and anticipation of the coming, looking back and looking forward, we are not able to see calculus as describing anything but correlation of individual abstract objects. More concretely, for accounting for the change of the position x of a physical object, calculus assigns to various values of x various values of a parameter t and correlates them according to well known rules. We do not view that as transitional change of x unless we already have brought in an idea thereof. Speaking of transitional change of some x, in turn, requires that x proceed along an abstract path t that accommodates an idea of moving, of becoming. To be sure, x may jump while proceeding along t, but t has to be a transitional path to accommodate the envisaged idea of transitional change. Having t be a set of individuated stages, even if those are totally ordered, even if arithmetically complete, would merely provide a set of parameters for x. But transition of x, proceeding of x through t, it would not be. If t is the set of real numbers, then proceeding from a given element t_1 of t—which we may refer to as "value"—to a prospective next one, t_2, is impossible, because, for that, some value t_{15}, with $t_1 < t_{15} < t_2$, would have to be passed, and for that, some value t_{12}, with $t_1 < t_{12} < t_{15}$, would have to be passed, and so on. Zeno's Achilles would never be able to take his first step, let alone to reach the end of the running track, we would merely have a plurality of mutually resembling copies of Achilles at various positions on the track, but Achilles would not run. No remedy is easily found, as long as t remains a set of individuals. Progression along t may be enabled by making t well-ordered. But even that works only for computer programs or similar procedures, but not for a transitional movement. For, between any two values t_1 and t_2 of t, with $t_1 < t_2$, we, having transition in mind, may insert another value t_{15}, with $t_1 < t_{15} < t_2$, through which the transition proceeds, and so on, which would destroy the well-ordering of t. On the other hand, refining real numbers by an arithmetic of infinitesimal numbers will not help, since the individuality of numbers is not broken (Blau 2016). There will still be nothing, but a—vastly extended—plurality of totally ordered static stages, but, as above, no transition, no change, no movement. If we, in general, aim to speak of becoming or of change of x along t, then it is hard to see how t can be but continuous. The required abstract continuous path t is what we call "time".

The preceding brief consideration arguing for continuously flowing time is not and cannot be made logically compelling. It hinges on an a priori intuition about the world undergoing change rather than being a plurality of frozen, static world stages, which is embedded already in infants and certainly also in many animals. Evidence of our preference of viewing the world as a changing entity is found in cognitive experiments, for example, those revealing the widely discussed phi movement phenomenon, a perceived movement of a flashing bright spot that is brought about by an alternately flashing pair of fixed lamps, even though a change of color is involved (Kolers and von Grünau 1976). The origin of that intuition cannot be traced back to an evolutionary advantage of having it. Whether a member of some species performs well for its own future or its offspring does not depend on it recognizing its future as the outcome of transitional flow. Good performance is taken care of by a natural process, which may as well come about in a series or tree structure of causally connected discrete worlds. But we take our intuition of us undergoing transition and transition being predicated on the flow of time serious by the following principle.

(P)   Strong intuitions we have which are not conducive to evolutionary success reflect, at least roughly, a metaphysical principle.

Here the metaphysical principle in question is the existence of transitional change, which, by the above, strongly suggests that time is continuous. An according stipulation of the continuity of time cannot be part of our formal theory ST, but has to precede it.

The way mental states are related to time may support what has just been said. Again, we can do no more than make a few simple observations from a topic that vastly exceeds the present scope. Upon introspection, mental states, if they are not independent of each other, appear to have temporal character. Mental states succeed each other, even though they partially overlap, and cause each other. They fade in and out and there is neither a way nor a need to make their temporal locations precise. But there is no spatial relation between mental states. Those observations underline what is obvious, a different character of time and space. In the usual set-up of spacetime as $\langle M, g \rangle$, the different character of time and space becomes manifest only in relatively high-level entities, such as the different signs of the lengths of timelike and spacelike tangent vectors. We propose a more intimate connection of the way mental states are mutually related and the basic constituents of spacetime. Already on a primitive level, time and space should differ. There are not many ways to express such a difference. Individuality and lack thereof is such a way. Primacy of time over space is another. We follow that cue, postulate time as primitve and continuous and construct from there spacetime as a higher-level entity in which continuity has been broken.

In the continuous flow of time, we cannot speak of individual time points. Rather, any given point in time, similar to a point in Euclidean geometry, is created on a line by having it intersected by another line. Intersection of two lines destroys their continuity. Yet adding more and more intersection points on a given line according to first order formulas results in a total order of densely arranged points, what is called "continuity" in the language of calculus. Yet it is no more than a weak and incomplete reflection of the original continuity.

Our construction of spacetime has proceeded accordingly, as described above in the second section. Time is continuous and hence structureless. We can conceive objects that represent time. They have to be structureless, as well, and so formally have to be individuals. In our set-up, time is represented by worldlines, the urelemente in the intended model of ST. Worldlines, as individuals, are capable of undergoing relations. The relation we have in mind is having a worldline w intersect another worldline v at some real number x. The set of real numbers at which a given v is intersected by other worldlines is totally ordered and weakly reflects the continuity of v. Now the construction proceeds as already described. Time is prior to spacetime. A fortiori, time is prior to space. Time is continuous, wheras spacetime is not.

Spacetime, lacking continuity, is faithfully mappable on an appropriate mathematical structure. Physical character of objects in our combined ontology $\mathbb{U}$ stems from having risen from worldlines, the urelemente in $\mathbb{U}$. In general, urelemente do not collapse into the empty set if they are associated with properties that go beyond the pure elementhood structure of $\mathbb{V}$. In the context of ST, the property in question is having continuity. Urelemente in ST are a set of individuals which are each continuous. The formally inexpressible property of continuity conveys physical character on the elements of the combined mathematico-spatiotemporal world.

5   Conclusion

In an axiomatic construction of spacetime by a first-order logic, ST, a conservative extension of Zermelo's Z, the wealth of relativistic spacetime is obtained from a tiny conceptual base of worldlines as urelemente and a ternary intersection relation. The model of ST is ontologically taken serious. There is no need to employ instrumental or ad hoc entities. Ontological primitives are rather conceived to be abstract, resting on ideas which we intuit and which look irreducible, individuality and continuity (as presently understood, see the appendix), without taking recourse to the concrete world.

The feasibiliy of such an approach has been demonstrated. Specifically, the ontological primitives, worldlines, are urelemente in an extended set-theoretical hierarchy. Being urelemente rather than the empty set, they carry a property which puts them beyond the pure element structure of the standard set-theoretical hierarchy. That property is continuity. It well reflects our intuition of the flow of time and is necessary for our view of a changing world.

Spacetime thus is established as an entity on its own, primary to physical objects such as particles and fields. Spacetime is physical, even though not subject to direct experience. The resulting ontology combines mathematical objects and a part of physics, in which the latter nevertheless is clearly distinguished.

Appendix: A note on terminology

It is understool to be risky, often even detrimental for effective scholarly communication, to deviate from standard usage of terminology. In the present article, such a step is nevertheless taken in two cases, the terms "worldline" as well as "continuity" and, with it, "continuous", since their standard usage is the outcome of presuppositions that are abandoned here. The present usage seems, in fact, to stay closer to an intuitive understanding of the words that is unencumbered by the history of calculus.

In our usage, formally, worldlines are individuals without elements. Worldline curves are curves in spacetime, those curves whose elements—spacetime points—contain a common worldline. Beyond the formal, in our usage, worldlines are continuous lines as defined below.

In standard usage, worldines are curves in spacetime, in particular, particle paths in spacetime. So, in standard usage, worldlines are sets of points. What we call "worldline curve" in our usage would be called "worldline" in standard usage.

In our usage, continuity is flow, becoming, transition, without individuated stages or steps or leaps. Continuity is exemplified by a line as we have it in mind when we draw it. Any attempt to analyze continuity in terms of individuated parts and relations thereof does not succeed. Continuity cannot be characterized formally. It cannot even be approximated formally. Continuity, thus understood, does not apply to sets or pluralities, but to (abstract) objects, such as naively conceived geometric lines. Time is taken to be continuous in the present approach.

In standard usage in the realm of logic and mathematics, continuity is a property of sets, characterizing a relation between elements thereof. For example, due to its arithmetical properties, the set of real numbers is said to be continuous and, in textbooks, a property of continuity is defined for functions.

References

[1] Andréka, H.; Madarász, J. X.; Németi, I. (2004). Logical analysis of relativity theories. In: Hendricks et al. (eds.), *First-order Logic Revisited*. Berlin: Logos Verlag, pp. 7-36.

[2] Andréka, H.; Madarász, J. X.; Németi I. (2006). Logical axiomatizations of spacetime; samples from the literature. In A. Prékopa and E. Molnár

(eds.), *Non-Euclidean Geometries*: János Bolyai Memorial Volume, pp. 155-185. New York: Springer.

[3] Andréka, H.; Madarász, J. X.; Németi I. (2007). Logic of space-time and relativity. In *Handbook of Spatial Logics*, pp. 607-711. New York: Springer.

[4] Andréka, H.; Madarász, J. X.; Németi, I. ; Székely, G. (2008). Axiomatizing relativistic dynamics without conservation postulates. *Studia Logica* 89, pp. 163-186.

[5] Andréka H., Madarász, J. X.; Németi I. and Székely G. (2012). A logic road from special relativity to general relativity, *Synthese* 186, pp. 633-649.

[6] Ax, J. (1978). The elementary foundations of spacetime. *Foundations of Physics*.8, pp. 507-546.

[7] Benda, Th. (2008). A formal construction of the spacetime manifold. *Journal of Philosophical Logic* 37, pp. 441-478.

[8] Benda, Th. (2015). An axiomatic foundation of relativistic spacetime. *Synthese* 192, pp. 2009-2024.

[9] Benda, Th. (2017). Basing the metric of relativistic spacetime on worldlines. To be submitted.

[10] Blau, U. (2008). *Die Logik der Unbestimmtheiten und Paradoxien*. In German. Heidelberg: Synchron Publishers.

[11] Blau, U. (2016). *Grundparadoxien, grenzenlose Arithmetik, Mystik*. In German. Heidelberg: Synchron Publishers.

[12] Brown, H., and Pooley, O. (2006). Minkowski Space-Time: A Glorious Non-Entity. In: D. Dieks (ed.), *The Ontology of Spacetime*, pp. 67-89. Amsterdam: Elsevier.

[13] Bunge, M. (1967). *Foundations of physics*. New York: Springer.

[14] Cambridge Dictionary (2013). *Cambridge Advanced Learner's Dictionary*. Cambridge University Press.

[15] Ehlers, J.; Pirani, R.A.E.; Schild, A. (1972). The geometry of free fall and light propagation. In: Raifeartaigh, L.O. (ed.), *General Relativity, Papers in Honour of J.L. Synge*, pp. 63-84. Oxford: Clarendon Press.

[16] French, S.; Krause, D. (2006). *Identity in Physics: A Historical, Philosophical, and Formal Analysis*. Oxford University Press.

[17] Gödel, K. (1964). What is Cantor's continuum problem? In: P. Benacerraf and H. Putnam (eds.), *Philosophy of Mathematics*, pp. 470-485. Cambridge University Press.

[18] Hawking, S.W.; King, A.R.; McCarthy, P.J. (1976). A new topology for curved space-time which incorporates the causal, differential, and conformal structures. *Journal of Mathematical Physics* 17, pp. 174-181.

[19] Kant I. (1998). *Critique of Pure Reason*, translated and edited by P. Guyer and A. Wood. New York: Cambridge University Press.

[20] Kolers P.A. and von Grünau, M. (1976). Shape and Color in Apparent Motion. *Vision Research* 16, pp. 329-35.

[21] Madarász, J. X.; Németi, I.; Székely, G. (2006). Twin Paradox and the logical foundation of relativity theory. *Foundations of Physics* 36, pp. 681-714. 33

[22] Madarász, J. X.; Németi, I.; Tőke, Cs. (2004). On generalizing the logic-approach to space-time towards general relativity: first steps. In: Hendricks et al. (eds), *First-order Logic Revisited*, Logos Verlag, Berlin, pp. 225-268.

[23] Malament, D. (1977a). The class of continuous timelike curves determines the topology of spacetime. *Journal of Mathematical Physics* 19, pp. 1399-1404.

[24] Malament, D. (1977b). Causal theories of time and the conventionality of simultaneity. *Nous* 11, pp. 293-300.

[25] Margenau H. (1944). The exclusion principle and its philosophical importance. *Philosophy of Science* 11, pp. 187-208.

[26] Mundy, B. (1986). Optical Axiomatization of Minkowski Space-Time Geometry *Philosophy of Science* 53, pp. 1-30.

[27] O'Neill, B. (1983). *Semi-Riemann Geometry*. San Diego: Academic Press.

[28] Quine, W.V. (1969). Ontological Relativity. Reprinted in: W.V. Quine, *Ontological Relativity and Other Essays*, pp. 26-68. New York: Columbia University Press.

[29] Reichenbach, H. (1924) *Axiomatization of the theory of relativity*. University of California Press, 1969.

[30] Robb A.A. (1936). *Geometry of Time and Space*. Cambridge: Cambridge University Press.

[31] Savellos, E. (1990). On defining identity. *Notre Dame Journal of Formal Logic* 31, pp. 476-484.

[32] Webster (2003). *Merriam-Webster's Collegiate Dictionary*.

12 A Copernican turn in temporal logics

PETR ŠVARNÝ

Abstract The article discusses the role of observers in perception of flow of time. It compares two established logics, Branching Space-times and Branching Continuations to a new logic based on Barbour's timeless approach to physics. The article shows that the introduction of observer based valuation allows for the same evaluation of statements in both temporal and atemporal models. We show this on the evaluation of statements about the future. Therefore we reach the conclusion that ontological time is not necessary for the evaluation of temporal statements.

1 Introduction

There is no time like the present. But what if every time is like the present? A scientific hypothesis or even a theory that goes against our daily experiences has it always difficult to convince people about its correctness. As a prominent example we can look at Copernicus and his challenge to the geocentric system. While the geocentric system accommodates our impression of the Sun's movement, it can correctly predict the positions of planets only if it uses a very complicated system of planetary paths. On the other hand, the heliocentric system presents a simpler tool for the prediction of planetary movements, nevertheless it forces us to challenge our daily perspective.

In a similar way we challenge our temporal perspective and the notion of flow of time. As in the case of Copernicus, we take into account the position of the observer in the universe. The view we partake to defend is that it is the specificity of observers and their position in time that leads to the perception of flow and the uniqueness of the present. The role of observers was often neglected and time was treated as a whole instead of being judged from the perspective of different observers. Following the recent contribution to the study of time by Dieks [4], we attempt to present a formal models for observer based temporal logics. We introduce a new logic

A. S. Stefanov, M. Giovanelli (Eds), *General Relativity 1916 - 2016. Selected peer-reviewed papers presented at the Fourth International Conference on the Nature and Ontology of Spacetime, dedicated to the 100th anniversary of the publication of General Relativity, 30 May - 2 June 2016, Golden Sands, Varna, Bulgaria* (Minkowski Institute Press, Montreal 2017). ISBN 978-1-927763-46-9 (softcover), ISBN 978-1-927763-47-6 (ebook).

based on Barbour's timeless physics and compare this with two established temporal logics, namely Branching spacetimes and Branching continuations. We discuss the role of observers in all of these systems and we investigate the truth and falsity of different temporal statements. Comparing a temporal and atemporal model allows us to demonstrate the weak Copernican principle in time - formally showing that the present, and its observers, do not need to be in any specially favoured position in the universe.

Firstly we discuss the philosophical and terminological foundations of the work, especially McTaggart's time series, Belnap's Branching spacetime, and Barbour's timeless Platonia. Thereafter we introduce the formal tools and logics based on these motivations and show how a timeless universe can seem to observers as containing time.

2   Philosophical background

A philosophical origin of this work can be found in [6]. McTaggart distinguishes three different time series called A, B, and C-series respectively. The A-series speaks about time as the 'future', 'present' and 'past' and hence also encompasses a privileged now and a dynamic flow of time. On the other hand, the B-series uses only the terms 'earlier than' and 'later than', thus speaks only about temporal relationship but not about any change. The last, C-series, is completely atemporal view and describes only the non-oriented relationship between events.

McTaggart argues that if we want to explain time and our experience of it, we need to look at the A-series. The C-series obviously does not describe our temporal experience, neither does the B-series. He concludes, however, that because the A-series is contradictory there cannot be any time itself. His argument against the A-series is not the aim of our paper, so let us just mention that it is far from definitive and was opposed by other authors. Nevertheless, the difference of the time series is a basic step in identifying basic perspectives on time.

An older discussion on this topic was already present in Greek philosophy, where we could say Parmenides defended that time is an illusion and Heraclitus argued for the opposite. A contemporary wording of their argument might be that according to Heraclitus "the world is made up of 3D objects, which endure and change in time, while retaining their identity from one moment to the next. Parmenideans, on the other hand, believe that the world is a changeless 4D spacetime continuum, containing material objects that are 4D worm-like volumes extended along the time dimension." [5].

A recent philosophical revisit of this problem can be found in [4]. Dieks argues for the B-series and concludes that "accounts of our experience of passage that rely on a postulated objective flow of time have not shown that they are more than abstract metaphysical exercises without a link to what science tells us about the world". He also points to the similarity with colours as their perception is not merely an illusion although there is no actual colour present. As Dieks sums it up: "In this sense our feeling of flow is veridical, in the same way as the perception of a colour can be faithful to an actual state of affairs."

Yet, also physics seem to support a world without a flow. Firstly, as mentioned in [4], general relativity and physics in general show that "our intuition of being in causal contact with a global Now is non-veridical". Secondly, we do not need time in physics at all. Barbour [1] presents physics that are completely timeless. Barbour builds up the world from so called 'configurations' and connects them with a specific measure. Although these configurations can be realized in multiple ways (for example as relative configurations of particles in Euclidean space), they form the 'primary ontological elements' for his theory. The measure then connects these configurations and gives the world a C-series-like form. This approach, based on physics, therefore replaces the classical linear idea of time with a multidimensional structure of possibilities.

These all, although well defended positions, represent non-formal approaches. However, we have at our disposal formal temporal logics that would allow us to formulate these positions in a formal way. Namely the branching temporal logic based on Belnap's work [3] attempts to capture relativistic spacetime. This temporal logic uses a structure composed of causally ordered point-events.[1] The higher order building blocks of these structures can vary among approaches from so called 'histories' in the original Branching Spacetimes [3] to the looser 'continuations' in Branching Continuations [8]. These logics are viewed by some as a possibility how to reconcile becoming with relativity [9]. We will only mention that there exists a specific type of models that bring BST even closer to physics, the so called Minkowski branching structures where every history is isomorphic with Minkowski spacetime [7][12]. Therefore even a demonstration closer to physics could be made.

The last notion that needs to be introduced is an observer. Notice that the previous philosophical account relied on the phenomenology of time and therefore it would rely the epistemic state of the observer. Thus in order to formalize the arguments we also need to use a similar observer. An observer is understood here as a local collection of measuring devices. For simplicity we will assume that the observer is not fallible and has all the available information at his disposal. This allows us to equate the observer's knowledge with the actual state that is physically accessible to the observer. The physical accessibility of the state of the world is limited by the fact that the observer is a local collection of measuring devices. An observer is therefore only a local collection of measuring devices with a history of measurements or in other words with a history of states. In the context of a spacetime, our definition confines the observer to a time-like finite worldline with all the events that can access this worldline via an at most light-like curve.[2]

---

[1] In the formal part of this paper, the term event might be used as instead of point-event for the sake of brevity.

[2] This definition of an observer might be non-standard as we do not use tetrads, or inextendible curves. A further analysis of the observer with regards to specific special relativistic or general relativistic properties with a more detailed physical account is planned in future work.

Before we start with temporal logics, a short note about the expression itself. We understand temporal logics as all kinds of logics that formalize our notions about time. Hence even an atemporal model like Barbour's would still be a temporal logic.

3 Branching Spacetimes with Observers

Let us sum up the basic ideas and definitions of Branching Spacetimes and present the role of observers. Although Branching spacetimes (BST) were introduced in [3], we present the concise version from [12].

Definition 1 (Def. 1 in [12]).

- *The set OW called* ***Our World*** *is composed of point-events e ordered by the causal relation $\leq$.*

- *A set $h \subseteq OW$ is* ***upward-directed*** *iff $\forall e_1, e_2 \in h \; \exists e \in h$ such that $e_1 \leq e$ and $e_2 \leq e$.*

- *A set h is* ***maximal with respect to the property of upward-directedness*** *iff $\forall g \in OW$ such that $h \subset g$, g is not upward-directed.*

- *A subset h of OW is a* ***history*** *iff it is a maximal upward-directed set.*

- *For histories h_1 and h_2, any maximal element in $h_1 \cap h_2$ is called a* ***choice point*** *for h_1 and h_2.*

Hence a history is close to the idea of a possible course of events. What might seem a little counter-intuitive is the scope of a history as it encompasses all the events of a possible course of the world. Two histories contain at least one distinct event, each history hence represents a different course of events in the universe. However, they are not separate ontological entities, because the events are the building blocks of Our World.

Definition 2 (BST model, Def. 2 in [12]). $\langle OW, \leq \rangle$ *where OW is a nonempty set and $\leq$ is a partial ordering on OW is a structure of BST iff it meets the following requirements:*

1. *The ordering $\leq$ is dense.*

2. *$\leq$ has no maximal elements.*

3. *Every lower bounded chain[3] in OW has an infimum in OW.*

4. *Every upper bounded chain in OW has a supremum in every history that contains it.*

5. *(Prior choice principle) For any lower bounded chain $O \subset h_1 - h_2$ there exists a point $e \in OW$ such that e is maximal in $h_1 \cap h_2$ and $\forall e' \in O(e < e')$.*

---

[3] A chain is a totally ordered set.

In order to investigate the truth of statements in a BST structures we should introduce also a language. This language contains classical logical operators and the usual Priorean operators F, P ('it will be true', 'it was true'.) The '*Sett :*' operator denotes a settled option, i.e. true for all histories. With these we can introduce the valuation of formulae.

Definition 3 (Point satisfies formula).

For the model $\mathfrak{M} = \langle OW, \leq, v \rangle$. *Where v is the valuation $v : Atoms \to \mathcal{P}(OW)$. For a given event e and history h, such that $e \in h$:*

$\mathfrak{M}, e, h \Vdash p$	iff $e \in v(p)$
$\mathfrak{M}, e, h \Vdash \neg\varphi$	iff not $\mathfrak{M}, e, h \Vdash \varphi$
$\mathfrak{M}, e, h \Vdash \varphi \wedge \psi$	iff $\mathfrak{M}, e, h \Vdash \varphi$ and $\mathfrak{M}, e, h \Vdash \psi$
$\mathfrak{M}, e, h \Vdash F\varphi$	iff there is $e' \in OW$ and $e^* \in h$ s.t.
	$e' \leq e^*$ and $\mathfrak{M}, e', h \Vdash \varphi$
$\mathfrak{M}, e, h \Vdash P\varphi$	iff there is an $e' \in h$ s.t. $e' \leq e$ and $\mathfrak{M}, c', h \Vdash \varphi$
$\mathfrak{M}, e, h \Vdash Sett : \varphi$	iff for all $e' \in h'$, for all h' such that $e \in h'$:
	$M, e', h' \Vdash \varphi$

Notice that the future operator could be rephrased as 'at a future event to e, namely e^*, we will be able to say that φ is true'. We cannot, however, just pick event e' as it does not have to be necessarily in h and it might be that φ does not hold at e^* (i.e. it is not a settled future)

This satisfaction definition is the classical version without an observer. We can, however, add an observer to BST using a few simple definitions and thereafter modify the evaluation of formulae.

Definition 4 (Worldline).

A set $Wl \subseteq OW$ is a worldline iff Wl is a chain. We denote $Wl_{(e)}$ a worldline containing the point-event e.

Definition 5 (Observer in BST).

*An **observer** $\mathcal{O}$ is a finite worldline Wl limited by two point-events e_i and e_f, where e_i is the initial observation and e_f is the final observation.*

*An **observer of a point-event** $e \in OW$, $\mathcal{O}_e$, is an observer such that there is $e' \in \mathcal{O}_e$ such that $e \leq e'$.*

*We define **observations of the observer** $\mathcal{O}$, $Obs_{\mathcal{O}}$, as the set of all point-events that were observed by observer $\mathcal{O}$.*

An observer hence can only observe point-events that have causally influenced his worldline. Because an observer is basically a set of consistent point-events, he does specify a set of histories that are consistent with each other up to the point of the last observer's point-event.

Definition 6 (Observer valuation).

For given $\mathfrak{M}, e$ and observer $\mathcal{O}_e$ and histories h_i, such that $\mathcal{O}_e \subset h_i$, a formula φ is true for $\mathcal{O}_e$ iff for every h_i it holds that $\mathfrak{M}, e, h_i \Vdash \varphi$.

Because of our approach, we take the observers worldline as the whole state of the observer based on which we can judge upon the truth or falsity of statements. We can thus state how a formula would be evaluated with respect to an observer.

Theorem 1 (Observer time asymmetry). *For any observer $\mathcal{O}$ that is part of at least two histories, there exists at least one well formed formula about the future that cannot be attributed any truth value.*

Proof. An observer's set of observations is part of at least two histories then these histories must coincide on the observer, they must diverge at some e such that $e_f \leq e$ and hence we can construct a formula that relies on the valuation at e that will have a different truth value in h_1 and h_2 and hence cannot be true or false for the observer. **Q.E.D.**

4 Branching Continuations with Observers

One of the original motives for Branching Continuations (BCont) in [8] were general relativistic spacetimes. The original BST was not capable of capturing general relativistic structure. Two properties of histories are problematic – their scale and their upward-directedness. However, we can introduce BCont with observers as in [11]. A noticeable difference at first sight between the two structures, a BST and a BCont one, is the locality of 'histories' in BCont. This brings them closer to the observer than in the case of BST histories as we have seen.

BCont also starts out with the set of point-events of Our World OW. However, these points are related by paths, called snake-links. However, if we follow the basic definitions from the mentioned articles, we do not need to make any alterations and BCont can accommodate the same type of observers as we have seen in BST.[4]

Definition 7 (Snake-link, Def. 1-3 in [8]).
The properties and basic definitions of snake-links:

1. *$\langle e_1, e_2, \ldots, e_n, \rangle \subseteq W \, (1 \leq n)$ is a snake-link iff*

$$\forall i : 0 < i < n \rightarrow (e_i \leq e_{i+1} \vee e_{i+1} \leq e_i)$$

2. *A snake-link is above (below) $e \in W$ if every element of it is strictly above (below) e.*

3. *Let $W' \subseteq W$ and $x, y \in W'$. x and y are snake-linked in W' iff there is a snake-link $\langle e_1, e_2, \ldots, e_n, \rangle$ such that such that $x = e_1$ and $y = e_n$ and $e_i \in W'$ for every $0 < i \leq n$.*

4. *For $x, y \in W$, x and y are snake-linked above e, $x \approx_e y$, iff there is a snake-link $\langle e_1, e_2, \ldots, e_n, \rangle$ above e such that $x = e_1$ and $y = e_n$.*

---

[4]For further discussion on the relation of BCont and BST, see [8].

The relation $\approx_e$ is reflexive, symmetrical and transitive, hence an equivalence relation on the set $W_e = \{e' \in W | e < e'\}$.

Definition 8 (Set of possible continuations, Def. 4 in [8]).
Set of possible continuations *of e, Π_e, is the partition of W_e induced by the relation $\approx_e$.*

$\forall e < x : \Pi_e \langle x \rangle$ *is the unique continuation of e to which the given x belongs.*

Fact 1 (Fact 5 in [8]).
$\forall e', e, e_0 \in W : ((e \le e' \vee e' \le e) \wedge e_0 < e \wedge e_0 < e' \to \exists H \in \Pi_{e_0} e, e' \in H)$

Definition 9 (Set CE of choice events, Def. 6 in [8]).
For $e \in W$, $e \in CE$ iff $card(\Pi_e) > 1$.

Definition 10 (Consistency, Def. 7 in [8]).
For $e, e' \in W$, let there be $W_e := x \in W | \forall c (c \in CE \wedge c < e \to c < x)$ and a similar for e'. Then e, e' are consistent *iff they are snake-linked within $W_e \cup W_{e'}$. A set $A \subseteq W$ is then* consistent *if every two elements of A are and it is* inconsistent *iff it is not consistent.*

Definition 11 (L-events, Def. 8 in [8]). $A \subseteq W$ *is an* l-event *iff $A \ne \emptyset$ and A is consistent.*

For the definition of a BCont model, the definition of a BST model is used, only altered on places, where the snake-link has its influence. BCont is in many aspects a generalised form of the BST models.

Definition 12 (Model of BCont, Def. 9 in [8]).
$\mathcal{W} = \langle W, \le \rangle$ *is a* model of BCont *if it satisfies:*

1. $\mathcal{W}$ *is a non-empty partially ordered set;*

2. *the ordering $\le$ is dense on W;*

3. W *has no maximal elements;*

4. *every lower bounded chain $C \subseteq W$ has an infimum;*

5. *if a chain $C \subseteq W$ is upper bounded and $C \le b$, then there is a unique minimum in $\{e \in W | C \le e \wedge e \le b\}$;*

6. *for every $x, y, e \in W$, if $e \not< x$ and $e \not< y$, then x and y are snake-linked in the subset $W_{e \not<} := \{e' \in W | e \not\le e'\}$ of W;*

7. *if $x, y \in W$ and $W_{\le xy} := \{e \in W | e \le x \wedge e \le y\} \ne \emptyset$, then $W_{\le xy}$ has a maximal element;*

8. *for every $x_1, x_2 \in W$, if $\forall c : c \in CE \to c \not< x_i$, then x_1, x_2 are snake-linked in the subset $W_{\not> CE} := \{e \in W | \forall c \in CE e \not> c\}$ of W.*

Further definitions enlighten, why are snake-links necessary. L-events are not as large as histories. For any two point-events from a history in BST, there is at worst an M-shaped path connecting them ordering wise. This cannot be guaranteed for BCont and l-events. For example, space-like related events cannot directly influence each other in a causal way. Therefore, they cannot be directly connected by $\leq$. If the l-event does not extend sufficiently to the future, it is the snake-link relation that allows us to connect the two point-events.

Definition 13 (Basic transitions in BCont, Def. 11 in [8]).
Let $\langle W, \leq \rangle$ be a model of BCont. A basic transition is a pair $\langle e, H \rangle$, where $e \in W$ and $H \in \Pi_e$ is a continuation of e.

Definition 14 (SLR, Def. 12 in [8]).
$e, e' \in W$ are SLR iff they are compatible but incomparable.

Another physics related term is S-t locations, which stands for spacetime locations.

Definition 15 (S-t locations, Def. 13 in [8]).
We say that a model $\langle W, \leq \rangle$ of BCont has spatio-temporal locations iff there is a partition S of W such that

1. *For each l-event A and each $s \in S$, the intersection $A \cap s$ contains at most one element;*

2. *S respects the ordering $\leq$, that is, for all l-events A, B, and all $s_1, s_2 \in S$, if all the intersections $A \cap s_1, A \cap s_2, B \cap s_1$ and $B \cap s_2$ are nonempty, and $A \cap s_1 = A \cap s_2$, then $B \cap s_1 = B \cap s_2$;*

3. *similarly for the strict ordering;*

4. *if $e_1 \leq e_2 \leq e_3$, then for every l-event A such that $s(e_1) \cap A \neq \emptyset$ and $s(e_3) \cap A \neq \emptyset$, there is an l-event A' such that $A \subseteq A'$ and $s(e_2) \cap A \neq \emptyset$, where $s(e_i)$ stands for a (unique) $s \in S$ such that $e_i \in s$;*

5. *if L is a chain of choice events in $\langle W, \leq \rangle$ upper bounded by e_0 and such that $\exists s \in S \ \forall x \in L \ \exists e \in W : (x < e \wedge s(e) = s)$, then $\exists e^* \left(e^* \in \bigcap_{x \in L} \Pi_x(e_0) = s \right)$.*

S is then called a set of s-t locations for $\langle W, \leq \rangle$.

Definition 16 (Ordering of s-t locations, Def. 14 in [8]).
For $s_1, s_2 \in S$, let $s_1 \precsim s_2$ iff $\exists e_1, e_2 (e_1 \in s_1 \wedge e_2 \in s_2 \wedge e_1 \leq e_2)$.

Lemma 1 (Fact 15 in [8]).
If $\langle W, \leq, S \rangle$, a BCont model with a set S of s-t locations, is downward directed, then $\precsim$ is a partial dense ordering on S.

Lemma 2 (Fact 16 in [8]).
Let $\langle W, \leq, S \rangle$ that is downward directed and satisfies the following conditions:

- $\forall e_1, e_2, c_3 \in W \; (e_1 \leq e_3 \wedge e_2 \leq e_3 \to e_1 \leq e_2 \vee e_2 \leq e_1)$ *surnamed "no backward forks"*

- $\forall e, e' \in W$: *if* e, e' *are incomparable by* $\leq$, *then there are* $H_1, H_2 \in \Pi_m$ *such that* $H_1 \neq H_2$, $e \in H_1$ *and* $e' \in H_2$, *where* m *is a maximal element of* $W_{\leq ee'} = \{y \leq e \wedge y \leq e'\}$;

Then S *is linearly ordered by* $\precsim$ *and every l-event of* $\langle W, \leq, S \rangle$ *is a chain.*

For semantics of BCont a point-event and l-event pair is used in a similar way as in BST. However, we use the definition from [8] used the original Branching Time models of Prior [10] as a basis for BCont semantics.

Definition 17 (BT+Instants inspired model, Def. 17 in [8]).
A model $\langle W, \leq, S \rangle$ *is said to be* (BT+Instants)-*like if it satisfies the following conditions:*

- *downward directedness,*

- *no backward forks,*

- $\forall e, e' \in W$: *if* e, e' *are incomparable by* $\leq$, *then there are* $H_1, H_2 \in \Pi_m$ *such that* $H_1 \neq H_2$, $e \in H_1$ *and* $e' \in H_2$, *where* m *is a maximal element of* $W_{\leq ee'} = \{y | y \leq e \wedge y \leq e'\}$;

This allows us to state the truth-conditions of metric tenses saying that the two events are t units apart. Sentences will be then judged based on evaluation points, built out of l-events and thus will be event/l-event pairs mentioned already earlier.

Definition 18 (Structure and model, Def. 18 in [8]).
A structure for the language $\mathcal{L}$, *as defined before, is a pair* $\mathfrak{G} = \langle \mathcal{W}, X \rangle$, *where* $\mathcal{W} = \langle W, \leq, S \rangle$ *is a (BT+Instants)-like model of BCont such that* $|S| = |\mathcal{R}|$, *and* X *is a real coordinalization of* S.
A pair $\mathfrak{M} = \langle \mathfrak{G}, \mathcal{I} \rangle$ *is a model for language* $\mathcal{L}$, *where* $\mathfrak{G}$ *is a structure for* $\mathcal{L}$ *and* $\mathcal{I} : Atoms \to \mathcal{P}(W)$ *is an interpretation function and Atoms is the set of atomic formulas of* $\mathcal{L}$.

Definition 19 (Evaluation points, Def. 19 in [8]).
Let $\mathfrak{G} = \langle \mathcal{W}, X \rangle$ *be a structure for language* $\mathcal{L}$, *where* $\mathcal{W} = \langle W, \leq, S \rangle$. *Then* $\langle e, A \rangle$, *written as* e/A, *is an* evaluation point *in* $\mathfrak{G}$ *for formulas of* $\mathcal{L}$ *iff* $\{e\} \cup A \subseteq W$ *and* $A \neq \emptyset$.

Noteworthy is the fact that we do not require for a e/A that $e \in A$, also to be mentioned is the fact that [8] suggests a plain ontological reading of the meaning of e/A. Although it is also true that the BCont approach carries with itself less tension between ontology and epistemology as l-events are more accessible than BST histories.

This construction of evaluation points and coordinalization of X allows us to use metric tense operators $F(x)$ and $P(x)$ with $x \in \mathbb{R}$. For the language $\mathcal{L}$, we assume that its atomic formulas are present-tensed and that it has the two metric tense operators, usual connectives ($\neg, \wedge, \vee, \to$) and modal operators $Sett$(as "it is settled"), $Poss$("it is possible") and an operator Now.

Definition 20 (Extensions of an evaluation point, Def. 20 in [8]).

Let $\mathfrak{G} = \langle \mathcal{W}, X \rangle$ be a structure for language $\mathcal{L}$, $\mathcal{W} = \langle W, \leq, S \rangle$, and e/A be an evaluation point in $\mathfrak{G}$ for $\mathcal{L}$. Then:

– e/A goes at least x-units-above e ($0 \leq x$) iff $\exists e_1 \in W \exists e_2 \in A(e_1 \leq e_2 \wedge int(e, e_1, x))$;

– e/A' is an x-units-above-e extension of e/A ($0 \leq x$) iff $A \subseteq A' \subseteq W$ and e/A' goes at least x-units-above e.

Definition 21 (Fan of evaluation points, Def. 22 in [8]).

Let $\mathfrak{G} = \langle \mathcal{W}, X \rangle$ be a structure for $\mathcal{L}$, $\mathcal{W} = \langle W, \leq, S \rangle$, and e/A be an evaluation point in $\mathfrak{G}$ for $\mathcal{L}$.

Two l-events A_1 and A_2 of $\mathcal{W}$ are isomorphic instant-wise iff $\forall e_1 \in A_1 \exists e_2 \in A_2 s(e_1) = s(e_2)$ and $\forall e_2 \in A_2 \exists e_1 \in A_1 s(e_1) = s(e_2)$

$e/A' \in \mathcal{F}_{e/A}$, fan of evaluation points determined by evaluation point e/A iff e/A' is an evaluation point in $\mathfrak{G}$ and A and A' are isomorphic instant-wise.

In many cases this leads to a single possible A', A itself. An important point is that the evaluation of the formula depends on the moment of use, e_C.

Definition 22 (Point fulfils formula, Def. 23 in [8]).

For given $e_C, e/A$ and the model $\mathfrak{M} = \langle \mathfrak{G}, \mathcal{I} \rangle$. Then:

1. if $\psi \in$ Atoms:$\mathfrak{M}, e_C, e/A \Vdash \psi$ iff $e \in \mathcal{I}(\phi)$;

2. if ψ is $\neg\varphi$: $\mathfrak{M}, e_C, e/A \Vdash \psi$ iff it is not the case that $\mathfrak{M}, e_C, e/A \Vdash \varphi$;

3. for $\wedge, \vee, \rightarrow$ also in the usual manner;

4. if ψ is $F_x\varphi$ for $x > 0$: $\mathfrak{M}, e_C, e/A \Vdash \psi$ iff there are $e' \in W$ and $e* \in A$ such that $e' \leq e*$ and int(e', e, x), and $\mathfrak{M}, e_C, e'/A \Vdash \varphi$;

5. if ψ is $P_x\varphi, x > 0$: $\mathfrak{M}, e_C, e/A \Vdash \psi$ iff there is $e' \in W$ such that $e' \cup A \in$ l-events and int(e',e,x) and $\mathfrak{M}, e_C, e'/A \Vdash \varphi$;

6. if ψ is Sett : φ : $\mathfrak{M}, e_C, e/A \Vdash \psi$ iff for every evaluation point e/A' from fan $\mathcal{F}_{e/A}$ and $\mathfrak{M}, e_C, e/A' \Vdash \varphi$;

7. Poss : $\psi := \neg$Sett : $\neg\psi$;

8. if ψ is Now : φ : $\mathfrak{M}, e_C, e/A \Vdash \psi$ iff there is $e' \in s(e_C)$ such that $e' \cup A \in$ l-events and $\mathfrak{M}, e_C, e/A' \Vdash \varphi$.

Definition 23 (Definite truth, Def. 24 in [8]).

$\mathfrak{M}, e_C, e/A \models \psi$, read as ψ is definitely true at $\mathfrak{M}, e_C, e/A$, iff there is an $x \geq 0$ such that for every x-units-above e extension e/A' of e/A : $\mathfrak{M}, e_C, e/A'f' \Vdash \psi$;

$\mathfrak{M}, e_C, e/A \models_{Indef} \psi$, read as ψ is indefinitely true at $\mathfrak{M}, e_C, e/A$, iff there is no $x \geq 0$ such that for every x-units-above e extension e/A' of e/A: $\mathfrak{M}, e_C, e/A' \Vdash \psi$ or for every x-units-above-e extension e/A' of e/A: $\mathfrak{M}, e_C, e/A' \Vdash \neg\psi$;

212

Lemma 3 (Thm. 25 in [8]). *For any formula ψ and any evaluation point e/A, exactly one of the following three options must hold: $e/A \models \psi$ or $e/A \models \neg\psi$ or $e/A \models_{Indef} \psi$*

We don't go into much more detail but let us list some of the properties from bcont.

– if ψ is fulfilled at an evaluation point e, it can cease to be fulfilled at an extension of this evaluation point;

– if ψ is definitely true at a evaluation point e, then it is definitely true in every extension of e;

– if ψ is indefinite at a point, so is its negation;

– if $\psi \wedge \varphi$ is indefinite at a point, $\psi \wedge \varphi$ is either indefinite or definitely false at this point;

– if $\psi \vee \varphi$ is indefinite at a point, $\psi \vee \varphi$ is either definitely true or indefinite at this point;

– if $\psi \to \varphi$ is indefinite at a point, $\psi \to \varphi$ is either definitely true or indefinite at this point;

– settled cannot be indefinite: $Sett : \psi$ is definitely true or $\neg Sett : \psi$ is definitely true.

Also in our coordinalization, every sentence becomes definitely true or definitely false at a sufficiently long extension of a initial evaluation point.

At this point we can use the same definitions of worldlines, observers and observer related truth as in the case of BST. Although we speak of consistent l-events instead of histories, the end result is the same and the BCont variation of Theorem 1 holds.

5 Barbourian Temporal Logic

Although this is a quote of Mach, it represents the key idea for Barbour's approach:

> It is utterly beyond our power to measure the changes of things by time... time is an abstraction at which we arrive by means of the changes of things; made because we are not restricted to any one definite measure, all being interconnected.
> [1]

As already mentioned, Barbour [1] authors what he calls a "many instants interpretation of quantum mechanics" a timeless model of the world where time is merely an abstraction. For Barbour, the equivalent of Our World is called Platonia and it is composed out of all the possible states of the world, sometimes referred to as the 'heap of possibilities'. One such possible state is called a 'configuration'.[5] For our purpose we do not need to go into much

---

[5]Configurations are sometimes referred to as instants. However, in order to purge any

213

detail about the quantum physical foundation of Barbour's theory. Let us just mention that the universe is represented by a Wheeler-DeWitt equation, a universe wave function that captures all the possibilities.

Hence in Barbour's view, each configuration is basically a three dimensional snapshot of the universe that captures the relative configuration of matter and fields in the universe. Platonia is then composed of all the possible configurations of such sort. They have an intrinsic structure that contains all the physical evidence that leads us to the impression that time passes. Barbour calls these objects 'time capsules'. A time capsule is therefore a part of the configuration that suggests in some way the direction of time or works as evidence for the passage of time. Classical examples of time capsules are geological sediments, camera films, or particle traces in a cloud chamber. However, as we see, each configuration is static and timeless.

We notice right away the difference between a BST or BCont structure and Platonia lies in the scale of their building blocks. While we had point-events at our disposal in the previous cases, Barbour's structure works with configurations. Configurations themselves have an intrinsic structure. However, for our purpose it is sufficient to take them as basic members of Platonia. This starts our work on Barbourian Temporal Logic (BTL).

Definition 24 (Platonia).
We call $\mathcal{P}$ Platonia, the set of all configurations c.

As mentioned, we assume a deeper structure in the configurations, for example time capsules are a part of configurations. We express this structure by one operator $\Delta(c, c')$ in our current approach. The meaning of $\Delta(c, c')$ is "the difference in arrangements" (i.e. relative distances, energies, etc.) between two configurations.[6]

This Δ has a significant impact on the way how we interpret and work with configurations. For one, this causes that configurations can be ordered and even that an analogue of "no backward branching" can hold in BTL.

Definition 25 (Direct transition).
Any two configurations $c, c' \in \mathcal{P}$ form a direct transition $c \approx_d c'$ iff $\forall c'' \in \mathcal{P} : \Delta(c, c') \leq \Delta(c, c'')$. There is a transition $c \approx c'$ iff there is a chain of direct transitions $c_1 \approx_d c_2 \approx_d ... \approx_d c_n$ such that $c_1 = c$ and $c_n = c'$.

Therefore if we would look at configurations of two points and have three possible configurations based on only the one dimensional distance of the points: c_1 one meter, c_2 two meters, and c_3 three meters, then there is a direct transition between c_1 and c_2, c_2 and c_3. However, there is not a direct transition between c_1 and c_3, because there exists a configuration whose arrangement is closer to the one of c_1, namely c_2. There still would be a transition between c_1 and c_3.

---

impression of temporality, we use the term configuration. Their meaning is the same in Barbour's theory.
   [6]A possible interpretation of Δ is the Kullback–Leibler divergence of configurations.

Definition 26 (Directly successive).

Two configurations $c, c' \in \mathcal{P}$ are directly successive $c < c'$ iff $c \approx_d c'$ and $c \in \Psi(c')$. Where $\Psi(c)$ denotes the set of possible preceding configurations based on time capsules from c.

Belnap [3] presents four possible ways of connecting events – causal dispersion, causal confluence, causal forward branching, and causal backward branching. The first, dispersion, connected one point-event to all point-events in its future that are causally influenced by it. These point-events had different spacetime coordinates (or has Belnap wrote the event "here-now" influences outcome "over-left-later" and "over-right-later"). Symmetrically for the past was defined the term causal dispersion. This relation cannot be directly translated into the language of configurations for it does not make sense to speak about a "here-now" configuration, because configurations are snapshots of the whole universe and hence cannot be differentiated based on the spatial dimension. Therefore causal dispersion and confluence of events cannot be captured in this framework. If two configurations c_1 and c_2 would be a causal dispersion of a configuration c_0, they would necessarily be equal to each other.

On the other hand, causal branching represented the relation between an event and its two possible outcomes (or causes). The relation between "here-now" and "possible outcome #1-here-later" and "possible outcome #2-here-later". This would already be possible to capture in terms of configurations. If we have two configurations c_1 and c_2 that capture the possible outcomes of an original configuration c_0, then both of them are directly successive to c_0 but different. This will lead later to the definition of choice configurations and a succession of configurations that are not branching will be a history.

However, we need to address backward branching. While Belnap in his original paper [3] mentions that backward "no backward branching is part of common sense" and does not delve too much deeper into the topic. Can we say something similar in the context of configurations? Namely that two incompatible configurations would lead to the same configuration? This should not happen thanks to time capsules. Basically our experience with time capsules, i.e. history, tells us that there is no backward branching. For if two different configurations c_1, c_2 would lead to the same outcome c_3, then there would be conflicting time capsules in c_3. One important note, however, is that we base our judgement on the relation of direct transition. There might exist some weird configuration c_w that would contain inconsistent time capsules and would pose as the result of both c_1 and c_2. Still, based on the definition of direct transition, we can assume that there is at least some configuration that has a smaller Δ to c_1, namely a configuration c_w without the inconsistent time capsules. Similarly for c_2.

Definition 27 (Barbour history).

A Barbour history h is a set of configurations $c_1, ..., c_n \in \mathcal{P}$ totally ordered by $<$.

Definition 28 (Choice configuration).

A choice configuration c_c is a configurations $c \in \mathcal{P}$ such that $\exists c_1, c_2 \in \mathcal{P}$: $c_1 \neq c_2$, $c < c_1$, $c < c_2$ and $c_c \in \Psi(c_1) \wedge c_c \in \Psi(c_2)$.

Definition 29 (Barbour structure).
A Barbour structure $\mathcal{S}$ given as $\langle \mathcal{P}, < \rangle$ satisfies the following:

1. *The ordering $<$ is dense.*

2. *The relation $<$ is transitive.*

3. *The relation $<$ is antisymmetric.*

4. *Every lower bounded chain in $\mathcal{P}$ has an infimum in $\mathcal{P}$.*

5. *Every upper bounded chain in $\mathcal{P}$ has a supremum in every history that contains it.*

6. *(PCP) For any lower bounded chain $C \in h_1 - h_2$ there exists a configuration $c \in \mathcal{P}$ such that c is maximal in $h_1 \cap h_2$ and $\forall c' \in C$ $c < c'$.*

The two structures, Barbour's and Belnap's, stay the same at this point. However, notice that in Barbour's idea of configuration ordering there could be a maximal element. A maximal element in this case can represent an ultimate arrangement of matter that does not have any successor. This would be also the trivial example of a Barbour structure that is not BST.

We use the language $\mathcal{L}$ with atomic formulas (statements about configurations in the present tense), tense operators F, P, modal operators $Sett :$, $Poss :$ and connectives: $\wedge, \vee, \rightarrow, \neg$. The semantic model itself needs only the addition of an interpretation $\mathcal{I} : Atom \rightarrow P(\mathcal{P})$. This interpretation is based on the time capsules of the configurations and their arrangements.

Definition 30 (Barbour model).
For the model $\mathfrak{M} =< \mathcal{S}, \mathcal{I}, \Vdash >$, a c from $\mathcal{P}$ satisfies a formula ψ in language $\mathcal{L}$ iff:

- $\psi \in Atom$: $\mathfrak{M}, c, h \Vdash \psi$ iff $c \in \mathcal{I}(\psi)$

- ψ is $\neg\phi$: $\mathfrak{M}, c, h \Vdash \psi$ iff it is not the case that $\mathfrak{M}, h \Vdash \phi$

- ψ is $\phi \wedge \pi$: $\mathfrak{M}, c, h \Vdash \psi$ iff $\mathfrak{M}, c, h \Vdash \phi$ and $\mathfrak{M}, c, h \Vdash \pi$

- ψ is $\phi \vee \pi$: $\mathfrak{M}, c, h \Vdash \psi$ iff $M, c, h \Vdash \phi$ or $M, c, h \Vdash \pi$

- ψ is $\phi \rightarrow \pi$: $\mathfrak{M}, c, h \Vdash \psi$ iff if $\mathfrak{M}, c, h \Vdash \phi$ then $\mathfrak{M}, c, h \Vdash \pi$

- ψ is $F\phi$: $\mathfrak{M}, c, h \Vdash \psi$ iff
  $\exists c' \in \mathcal{P} : c < c'$ and $\exists h' \subset \mathcal{P} : c, c' \in h'$ and $\mathfrak{M}, c', h' \Vdash \phi$

- ψ is $P\phi$: $\mathfrak{M}, c, h \Vdash \psi$ iff
  $\exists c' \in \mathcal{P} : c' < c$ and $\mathfrak{M}, c', h \Vdash \phi$

- ψ is $Sett : \phi$: $\mathfrak{M}, c, h \Vdash \psi$ iff
  $\forall h' \subset \mathcal{P} \forall c' \in \mathcal{P} :$ if $c \in h'$ and $(c' < c$ or $c < c')$ then $\mathfrak{M}, c', h' \Vdash \phi$

- ψ is $Poss : \phi$: $\mathfrak{M}, c, h \Vdash \psi$ iff
  $\mathfrak{M}, c, h \Vdash \neg Sett : \neg\phi$

Similarly as in the previous cases, we can introduce an observer. BTL cannot properly use the notion of a worldline. However, we can define the observer based on the configurations that have the closest to his current configuration. Notice that any (reasonable) configuration that contains an observer has also his history in the form of a time-capsule. Hence we can say that we tie together the the configurations based on the time-capsules available to the observer.

Definition 31 (Evidence).

A set $E \subseteq \mathcal{P}$ is called evidence iff E is a chain of configurations. We denote $E_{(c)}$ an evidence containing the configuration c.

Evidence is available at configuration c iff the set E contains c as the maximal member.

Note that we have definitely transitioned now from ontology to epistemics because the evidence present at a configuration c is some physical evidence in the configuration, however the chain of configurations that was created is just an abstraction based on Δ. Also do not forget that we assumed our observers are infallible and have all the accessible data available. This simplification, put into Platonia, would actually mean we single out some specific configurations (namely the ones containing such observers) of the plethora of possibilities.

Definition 32 (Observer in BTL).

An observer at configuration c, $\mathcal{O}_c$ is an observer that can use only the evidence available to him at c.

Definition 33 (Observer valuation in BTL).

For given $\mathfrak{M}, c$ and observer $\mathcal{O}_c$ and evidences h_i, such that $E \subseteq h_i$, a formula φ is true for $\mathcal{O}_e$ iff for every h_i it holds that $\mathfrak{M}, e, h_i \Vdash \varphi$.

Theorem 2 (Observer time asymmetry in BTL). *For any observer $\mathcal{O}_c$, if he is part of at least two histories, there exists at least one well-formed formula about the future that he cannot attribute any truth value.*

Proof. Similar as previously in Theorem 1. **Q.E.D.**

Therefore we see that also observers in BTL are subject to the same time asymmetry as observers in BST or BCont and could not, based on statements about the future, differentiate between a temporal and an atemporal model.

6 Conclusion

We presented three different temporal logics - Branching spacetimes, Branching continuations and Belnapian temporal logic. We hope to have shown that from an observers perspective it can be difficult to differentiate between there three models, although two of the temporal logics contain time in the form of spacetime and the third one is based on an atemporal universe. The evaluation of formulae for an observer can lead to the same models.

A further investigation on precise formulae that would allow to differentiate the structures should be now conducted in order to strengthen or refute the view that these models are equivalent from the point of view of an observer. However, there is a distinct possibility that the easier model to explain the world around us is atemporal. Nevertheless, as the Sun did not stop to rise in the east after people realized Earth is not the centre of the universe, so change does not vanish with the realization that the world might be timeless.

References

[1] Barbour, Julian B., The End of Time: The Next Revolution in Physics, 2000, Oxford University Press.

[2] Barbour, Julian B., The Nature of Time, arXiv preprint arXiv:0903.3489, 2009.

[3] Belnap, Nuel, Branching Space-Time, 1992, Synthese, 92, 3, 385–434.

[4] Dieks, Dennis, Physical Time and Experienced Time, 2016, Cosmological and Psychological Time, 3–20.

[5] McCall, Storrs, Philosophical consequences of the twins paradox, Philosophy and Foundations of Physics, 1, 191–204, 2006, Elsevier.

[6] McTaggart, John M. E., The Unreality of Time, 1908, Mind, 17, 68, 457–474.

[7] Müller, Thomas, Branching space-time, modal logic and the counterfactual conditional, Non-locality and modality, 273–291, 2002, Springer.

[8] Placek, Tomasz, Possibilities without possible worlds/histories, 2011, Journal of Philosophical Logic, 40, 6, 737–765.

[9] Pooley, Oliver, XVI—Relativity, the Open Future, and the Passage of Time, Proceedings of the Aristotelian Society, 113, 3 pt 3, 321–363, 2013, The Oxford University Press.

[10] Prior, Arthur N. Papers on Time and Tense, 1968, Oxford University Press.

[11] Švarný, Petr, Flow of Time in BST/BCONT Models and Related Semantical Observations, The Logica Yearbook 2012, 199–218, 2013, College publications.

[12] Wroński, Leszek and Placek, Tomasz, On Minkowskian branching structures, Studies In History and Philosophy of Science Part B: Studies In History and Philosophy of Modern Physics, 40, 3, 251–258, 2009, Elsevier.

Part IV

SPACETIME IN LINGUISTIC CONTEXT

A. S. Stefanov, M. Giovanelli (Eds), *General Relativity 1916 - 2016. Selected peer-reviewed papers presented at the Fourth International Conference on the Nature and Ontology of Spacetime, dedicated to the 100th anniversary of the publication of General Relativity, 30 May - 2 June 2016, Golden Sands, Varna, Bulgaria* (Minkowski Institute Press, Montreal 2017). ISBN 978-1-927763-46-9 (softcover), ISBN 978-1-927763-47-6 (ebook).

13 SPACETIME IN LANGUAGE

FEDERICO SILVAGNI

Abstract This paper is devoted to inner aspect (i.e. the spatiotemporal information encoded in words) and its relation to spacetime. It constitutes a first attempt to answer basic questions of linguistics by
taking advantage of the science and philosophy of spacetime. The
paper presents a long-standing problem in linguistics (namely, the
definition and the study of event predicates) and illustrates how it
can be addressed by bringing to linguistics the idea of events as
spacetime points that arises from the 4D understanding of physical
world. I propose that the aspectual primitive of event predicates
is a spacetime point and I show that this approach constitutes a
new and powerful insight for linguistics, while paving the way for
a promising connection between cognitive science and the fields of
physics and philosophy of spacetime.

1   Introduction

The fact that natural languages encode and convey spatial and temporal
notions is one of the most ancient and influential philosophical reflections
in linguistics. This idea finds its origins in Aristotle's Metaphysics and, at
present, represents a very well established knowledge of linguistic research,
which has been labelled as "aspect" [1–3].

Nowadays, linguists agree that aspect comes in two forms:

- Spatiotemporal notions that are encoded in predicates, that is to say, in
  words.[1] This information depends on the kind of eventuality[2] described

---

[1] In linguistics, there are many ways to understand the label *predicate*. We can isolate at
least two: the logical one defines a predicate as a function that takes at least an argument
and assigns a thematic role to it, and the general one, that defines a predicate as a word
endowed with some information of the world. Under the second definition, any word that
denotes some concept of the world can be considered a predicate. For the purposes of this
text, the second notion can be taken into account.

[2] The term *eventuality* is borrowed from Bach [10] and is used to make reference to any
possible circumstance denoted by a predicate, no matter its aspectual identity. It is, in
other words, a neutral label to make reference to any predicate.

A. S. Stefanov, M. Giovanelli (Eds), *General Relativity 1916 - 2016. Selected
peer-reviewed papers presented at the Fourth International Conference on the Nature and
Ontology of Spacetime, dedicated to the 100th anniversary of the publication of General
Relativity, 30 May - 2 June 2016, Golden Sands, Varna, Bulgaria* (Minkowski Institute
Press, Montreal 2017). ISBN 978-1-927763-46-9 (softcover), ISBN 978-1-927763-47-6
(ebook).

by a specific predicate and is referred to as "inner aspect" (i.e. inner with respect to words).

- Spatiotemporal notions that are not encoded in words, but are conveyed by grammatical categories such as tense and grammatical aspect.[3] This content informs about when, where and how a specific eventuality takes place in the world and is known as "outer aspect" (i.e. outer with respect to words).[4]

In order to grasp better the distinction, let's take the example in (1).

(1) Mary ran.

Leaving aside the nominal phrase (*Mary*), and taking into account the verbal element (*to run*), we can observe that this verb denotes a dynamic event, that is to say, an event that consists of "successive phases following one another in time" [1, p. 22]. Such a "dynamism" would thus be considered the inner aspectual information of the verb *run*. Moreover, the verb also carries a past tense inflexion, which informs that the event took place in the world in the past, that is to say, in a moment which is previous to the utterance of the sentence. Such a temporal information does not make up the identity of *run* (as inner aspect does), but serves as a temporal location of the event in the real world, and therefore belongs to an aspectual level that is "outer" to words.

This paper is dedicated to inner aspect. The research on inner aspect has always been focused on detecting a set of "primitive"[5] spatiotemporal notions that are responsible for the inner aspectual makeup of predicates and their syntactic and semantic effects. Regardless of the specific approach, linguists generally agree on the existence of, at least, two primitives: dynamism (which was referred to above) and telicity. On the one hand, the concept of dynamism derives from the Aristotelian concept of "kinesis", and it is understood as a movement, process, or change in time.[6] On the other hand, telicity (from the Greek "telos", which means "end" or "goal") is understood as the property of a predicate denoting an event that tends toward a limit.

Depending on the inventory of aspectual primitives encoded in predicates, linguists usually classify words into different eventuality-classes. The traditional and the most recognized taxonomy is borrowed from Vendler [1] and distinguishes among, at least, three eventuality-types: non-dynamic predi-

---

[3]Deictics such as *here, there*, etc. could also be considered part of the "outer" aspectual system, because they refer to specific locations in the real world where the main eventuality takes place.

[4]Strictly speaking, "outer aspect" is used in linguistics as a synonym of "grammatical aspect" (i.e. imperfective, perfective, etc.). Nevertheless, both tense (present, past or future) and deictics can be considered as "outer" aspectual elements because they are not primitives encoded in words and they convey spatial and temporal notions that locate eventualities in reality.

[5]In linguistics, a "primitive" is understood as the smallest unit (or "building block") of a specific content; in the case at stake, the inner aspectual content of predicates.

[6]See Acedo Moreno [27] on the concept of "kinesis", and the classical definitions of dynamism by Vendler [1], Comrie [3] and Kenny [2].

cates (which are also and exclusively atelic); dynamic and atelic predicates; and dynamic and telic predicates. The taxonomy is depicted in (Figure 1).[7]

[- dynamic]	[+ dynamic]	
(to be) {intelligent / tall / sick / tired / angry}, to love, to know, to sit, to stand, to lie...	[- telic]	[+ telic]
	to run, to walk, to eat...	*to break, to reach...*

Figure 1. Traditional aspectual taxonomy of predicates

The two classes of non-dynamic and dynamic predicates have been traditionally referred to as "States" and "Events", respectively. The label "State" refers, evidently, to the static nature of non-dynamic predicates, while the label "Event" has been assigned based on the widespread idea that events are processual, i.e. dynamic, in nature.

Leaving aside telicity, which is not relevant for the State / Event distinction (due to the fact that only dynamic predicates can be telics), as we will see below, the study of Events as dynamic predicates raises a huge problem, which was originally observed by Dowty [4] and presently remains unsolved.

This contribution addresses such a problem and constitutes a first attempt to derive a solution for it, by taking advantage of the concept of "spacetime". The paper will be broken down into the following sections: in the next section (section 2), I introduce the dilemma of the State / Event distinction; in section three, I present my proposal and I show the benefits associated with it; finally, in section 4, I draw my conclusions, as well as highlighting the new research opportunities that a connection between language and the concept of spacetime can give rise to. I wish to remind the reader that this is only a first exploration of the possible benefits that the concept of spacetime can bring to the study of the language faculty; likewise, despite my efforts to make linguistic concepts as accessible as possible, I would like to apologize in advance to any non-linguist reader for any difficulties experienced throughout the text and for any scientific misunderstandings on the part of the linguist author.

2   Between States and Events: the problem

As stated above, the philosophical judgment that lies at the basis of the study of inner aspect relates the notion of eventivity (or "happening") to the notion of dynamism.[8] According to this assumption, predicates are divided into Events and, by extension, States, depending on their (non-)dynamic nature. Some examples of eventive and stative predicates are given in (2).

---

[7]Vendler [1] makes use of a third primitive: duration. Nevertheless, it is not a well established fact that duration is an inner aspectual primitive, and many linguists reject its relevance [28–30]. Here, I show only the general view in linguistics, based on dynamism and telicity.

[8]Occasionally, dynamism, in turn, has been related to the notion of 'action' [31].

(2) States [non-dynamic]: *(to be) tall, to love, to know...*
     Events [dynamic]: *to run, to walk, to eat, to read...*

Linguists base the distinction on a set of tests that serve as evidence for the presence of a specific aspectual content in predicates.[9] Regarding the State / Event contrast, the traditional test is the progressive construction (*to be -ing*): only Events can appear in the progressive (3b), while States are excluded from this structure (3a) [1,2,5].

(3) a. *Ana is knowing French.[10]
     b. Ana is walking.

Beside this, many other tests distinguish between the two classes. Syntactically, only Events allow the adjunction of secondary non-selected predicates, as shown in (4b). Likewise, only Events can function as secondary non-selected predicates (5b), they allow event-related modifiers (namely, locatives and comitatives) (6b), they can appear in the complement position of perception verbs (7b) and they can be quantified (8b). Oppositely, States are refractory to all these syntactic phenomena: they do not allow for nor function as secondary non-selected predicates (4-5a), they exclude event-related modifiers (6a), they cannot appear in perceptual reports (7a) and they cannot be quantified (8a).[11]

(4) a. *Anna *loves music* tired.
     b. Ana *went* to school very tired.

(5) a. *Anna went to work *French*.
     b. John painted the table (while) *talking to his brother*.

(6) a. *Anna *knows French* {at home / with her brother}.
     b. John *read* the newspaper {at home / with her brother}.

(7) a. *I saw Anna *loving music*.
     b. I saw Anna *reading* the book.

(8) a. *Whenever Ana *knows French*...
     b. Whenever Ana *talks* to her brother...

Semantically, Events, as such, are interpreted as "happenings". In those languages in which the present tense can receive an actual reading (that is to say, it can be read as a syncretic form of the *to be -ing* construction), when Events are uttered in the present tense, they can be interpreted as happening "here and now", at the time of the utterance.[12] This is the case of

---

[9]The tests, which serve as a basis for detecting aspectual primitives and delimit different classes of predicates, are, at the same time, the phenomena that linguists try to explain by means of specific theories.

[10]In linguistics, the asterisk (*) marks agrammaticality. This way, we can represent and take into account minimal pairs in which one data is right and the other agrammatical.

[11]In the examples from (4) to (8), italics mark the predicate under analysis and the underlined part of the sentence coincides with the syntactic phenomenon at stake.

[12]The "time of the utterance" is a very well established notion in linguistics and it refers

Romance languages, for example, as shown in (9b).[13] Likewise, in the past (or future) tense, Events can be interpreted as happening once or several times (10b). On the contrary, States are read as generic statements, that is to say, as a property of the subject, and not as something that is happening, has happened or will happen (9-10a).

(9) a.1 Anne adore la musique classique. (#Ici et maintenant)[14] (French)
     a.2 Anna adora la musica classica. (#Qui e ora) (Italian)
     a.3 Ana adora a música clássica. (#Aqui e agora) (Portuguese)
     a.4 Ana adora la música clásica. (#Aquí y ahora) (Spanish)
         'Anna loves classical music. (Here and now)'[15]

     b.1 Anne parle au téléphone. (Ici et maintenant) (French)
     b.2 Anna parla al telefono. (Qui e ora) (Italian)
     b.3 Ana fala pelo telefone. (Aqui e agora) (Portuguese)
     b.4 Ana habla por teléfono. (Aquí y ahora) (Spanish)
         'Anna talks on the telephone. (Here and now)'

(10) a. As a child, she loved classical music. (#Only once/#Several times)
      b. During our last trip, Anna spoke to me (only once / several times)

All together, these tests make up a diagnostic of eventivity that allows linguists to classify predicates and to formulate theories that can explain their linguistic behaviour. Regarding the aspectual primitive which underlies the syntactic phenomena of the diagnostic (4-8) and the semantic interpretation of predicates as Events (9-10), we have already observed that, following philosophers such as Vendler [1], Kenny [2], Comrie [3], linguists have generally attributed the contrasts observed in the tests to the (non-)dynamic nature of the predicate: dynamism is the aspectual primitive that is responsible for the syntactic and semantic behaviour of a predicate as an Event.

Nevertheless, languages have many predicates that do not describe dynamic eventualities but behave as Events, both syntactically and semantically, as shown in examples (11)-(18). More concretely, there are many predicates that are static, but can appear in the progressive construction (11), allow and function as secondary non-selected predicates (12)-(13), can be modified by event-related modifiers (14), can appear in perceptual reports (15), can be quantified (16), and are read as happening at the time of the utterance, once or several times (17)-(18). This is the case of static verbs

---

to the moment at which a sentence is verbalized. The term *utterance*, in turn, can be understood as the act of verbalizing a sentence.

[13] This reading cannot be tested in English, because in English the present tense lacks the actual (*to be -ing*) reading. Nevertheless, the fact that only Events, contrary to States, can participate in the *to be -ing* construction is a syntactic counterpart of the interpretive phenomenon we are testing in (9) and it has already been observed for English in (3).

[14] In linguistics, the hashtag sign (#) is used to indicate that an utterance is pragmatically inadequate. In this case, for example, it marks that "here and now" is not a suitable reading for the sentence.

[15] English sentences between single quotation marks are literal translations for the reader. Whether they are grammatical or not, pragmatically adequate or not, does not matter, because they are not the focus of the analysis.

such as *to sit, to stand, to lie, to hang,* etc. and non-verbal predicates such as *tired, angry, hungry,* etc.[16,17]

(11) a. The socks are *lying* under the bed.
      b. Your glass is *sitting* near the edge of the table.
      c. The long box is *standing* on end.

[4, p.173]

(12) a. Anna *sits* motionless.
      b. John *lies* sick.

(13) a. Anna went to work very *tired.*
      b. John talked to his brother *lying on his bed.*

(14) a. The pearls *gleamed* in her hair. [6, p.819-820]
      b. Anna *sits* at the door of her house.

(15) a. I saw the child *sit* on the bench. [6, p.819]
      b. I saw John very *tired.*

(16) a. Whenever Anna *sits* in the kitchen...
      b. Whenever Anna is *tired*...

(17) a. Anne est assise à la table. (Ici et maintenant)   (French)
      b. Anna è seduta a tavola. (Qui e ora)   (Italian)
      c. Ana está sentada à mesa. (Aqui e agora)   (Portuguese)
      d. Ana está sentada en la mesa. (Aquí y ahora)   (Spanish)
      'Anna sits at the table. (Here and now)'

(18) a. During our last trip, Anna sat at the table (only once / several times).
      b. During our last trip, Anna was alone (only once / several times).

Contrary to the traditional view, which related the notion of "happening" to the notion of "dynamism", from these data we observe that languages have many predicates that describe "happenings", "situations", that is to say, Events, without being dynamic; in other words, as odd as it may seems according to the canonical assumptions, static Events exist.

The fact that some static predicates behave as Events was originally observed by Dowty [4] and it has recently been discussed by Maienborn [6–8]. In light of this evidence, linguists have not questioned the concept of Event as a dynamic happening, but they have limited themselves to treat those "static predicates that behave as Events" as nothing more than a special, unconventional, borderline class of States (see, for example, the taxonomies in [4,9–13], among many others). Nevertheless, it is clear that what the data

---

[16]The eventive nature of this specific group of non-verbal predicates (which are also known as "Stage-Level Predicates" in linguistics) has been thoroughly addressed in Silvagni [21,22,32,23].

[17]Again, in examples (11)-(16), italics mark the predicate under analysis, while the underlying stresses the syntactic phenomenon at stake.

suggest is something stronger than a simple exception in language, that is to say, a real problem: the fact that we can delimit a wide group of predicates that behave as Events (both syntactically and semantically) although they are not dynamic suggests that, contrary to the traditional assumption, dynamism is not a sufficient defining primitive of Events, and that at the basis of eventivity there is another aspectual content at play. In other words, it is evident from the data in (11)-(18) that the syntactic phenomena of the diagnostic and the interpretation of predicates as Events depend on an aspectual content that is not dynamism. This problem raises the fundamental research question in (19).

(19)   What is the aspectual primitive of eventivity (if it is not dynamism)?

3   Proposal

Following the traditional (dynamism-based) approach to eventivity, we saw that predicates that describe dynamic happenings are studied as Events, and any other predicate that describes a non-dynamic eventuality is studied as a non-Event (or State). According to this point of view, both predicates that describe properties of entities (which are also known as Individual-Level Predicates [14,15]) and those predicates that describe non-dynamic happenings, or situations (which are also known as Stage-Level Predicates [14,15]) fall under the class of States. The taxonomy is depicted in (Figure 2).

States		Events [dynamism]
Property-descriptive:	**Happening-descriptive:**	
(to be) {tall / intelligent}, to love, to know...	*(to be) {sick / angry / hungry / tired}, to sit, to stand, to lie...*	*to run, to walk, to eat, to read...*

Figure 2. Dynamism-based taxonomy

As we have just observed, this approach is not satisfactory, because it does not provide a correct understanding of eventivity, neither from a philosophical-ontological and conceptual point of view, nor from a merely linguistic perspective. This is because, on the one hand, from a conceptual point of view, happening-descriptive predicates (which are considered non-Events following the dynamism-based account of eventivity) are interpreted, contrary to predictions, as events. On the other hand, from a linguistic (syntactic) point of view, happening-descriptive predicates show the same linguistic behaviour as dynamic events.

We need, thus, a more basic primitive, which might be (roughly speaking) "smaller" than dynamism and, hence, capable to account for every eventive predicate, regardless of its dynamic or non-dynamic nature. (Dynamism would be consequently understood as a subordinate aspectual primitive — the same way telicity is subordinate to dynamism, for example, (Figure 1)—, and would be responsible for distinguishing among different types of events, without being the defining characteristic of eventivity as a whole).

I suggest that an answer to the problem can be found in the notion of "event" that stems from the concept of "spacetime", that is to say, from the discovery derived from Einstein's relativity theory that space and time are interwoven dimensions [16–19].[18]

As the mathematician and the philosopher comprehend with greater ease than the linguist,[19] under the 4D account of reality promoted by Einstein's and Minkowski's works, the world is understood as a continuous manifold of spacetime points, that is to say, realized or potential events [20, p. 51]:

> The world of physical phenomena which was briefly called "world" by Minkowski is naturally four dimensional in the space-time sense. For it is composed of individual events, each of which is described by four numbers, namely, three space coordinates x, y, z, and a time co-ordinate, the time value t. The "world" is in this sense also a continuum; for to every event there are as many "neighbouring" events (realised or at least thinkable) as we care to choose, the co-ordinates x_1, y_1, z_1, t_1 of which differ by an indefinitely small amount from those of the event x, y, z, t originally considered.[20]

I propose to borrow such a concept of event as a spacetime point in the four-dimensional continuum [20, p. 78] and transfer it to linguistics. Concretely, as in the world of physical phenomena individual events are spacetime points, I put forward the idea that events are conceptualized and encoded in language in the same way, as spacetime points.

Moving the notion of event away from the idea of dynamism, and relating eventivity to spacetime points represents a direct solution of the linguistic problem discussed in this paper. Under this view, in fact, any predicate that describes an happening has to be considered an event, no matter whether it is dynamic or static. In fact, if it is a point (and not a sequence of points)[21] that defines the conceptual identity of an event, it follows that a static happening

---

[18] In this regard, we can read in Einstein's words that "the four-dimensional mode of consideration of the world is natural on the theory of relativity, since according to this theory time is robbed of its independence" [20, p. 52]; "according to the theory of relativity, the time x_4, enters into natural laws in the same form as the space co-ordinates x_1, x_2, x_3" [20, p.104].

[19] Einstein's words on this topic are really evocative: "the non-mathematician is seized by a mysterious shuddering when he hears of "fourdimensional" things, by a feeling not unlike that awakened by thoughts of the occult" [20, p. 51].

[20] See also Minkowski's mathematical model of spacetime [33] and Einstein's observations on similarity to Minkowki's work [34].

[21] Note that the notion of "spacetime point" can also account for the notion of "dynamism", which traditionally has not received a clear definition, despite generally being accepted as a relevant aspectual primitive. Following the proposal of relating eventivity to the notion of "spacetime point", dynamism could be understood as a sequence of spacetime points. Moreover, this vision would be coherent with the evidence we are observing that dynamism is a subordinate property of eventivity.

(that is to say, an event which is conceived of as an unaltered, stationary, situation in spacetime) is also an event.[22]

Under these assumptions, I want to propose that the aspectual primitive encoded in event predicates is a spacetime point; consequently, the answer to the research question in (19) would be as in (20).

(20)   The primitive of eventivity is a spacetime point.

The idea is that eventive predicates (or Events) encode an aspectual primitive which is understood as a spacetime point, while non-eventive predicates (or States) do not encode any aspectual primitive, which means that they are devoid of any aspectual (i.e. spatiotemporal) notion. As outlined above, this approach leads to a more coherent understanding of inner aspect. In fact, it predicts that those predicates that do not describe happenings, that is to say, property-descriptive predicates, lack inner aspect and hence behave as non-Events linguistically. Likewise, it predicts that predicates that denote happenings, regardless of their static or dynamic understanding, do encode inner aspect, which gives rise to their linguistics behaviour as Events.Therefore, the present account of eventivity as a spacetime point (not as dynamism) solves the problem stated above, and justifies a taxonomy of predicates that is faithful to linguistic evidence, as depicted in (Figure 3). Now, dynamism is no more a relevant aspectual primitive (at least in the make up of eventivity), and if we were to consider it, it would occupy a subordinate status within Events. That is what we were looking for.

States (property-descriptive)	Events (happening-descriptive) [spacetime point]	
	(Non-dynamic)	(Dynamic)
(to be) {tall / intelligent}, to love, to know...	*(to be) {sick / angry / hungry / tired}, to sit, to stand, to lie...*	*to run, to walk, to eat, to read...*

Figure 3. Spacetime point-based taxonomy

At this point, maybe, some reader (and specially the non-linguist one) might be wondering about what it means for a predicate to encode a spacetime point. A clarification is necessary on this issue. Once we leave the

---

[22]Bear in mind that when talking about inner aspect we are not taking into account the duration of events. Inner aspect makes up the conceptual identity of a predicate and determines its syntactic behaviour, while duration is an outer aspectual concept, which depends on the specific realization of an event in the real world. The "spacetime point" primitive we are talking about in this paper has to be understood linguistically as an abstract aspectual concept, which determines the understanding of a predicate as an event. It should not be understood as a discrete point in the real world, because, obviously, any event has a specific duration and therefore coincides with a number of points in the physical world. It is outer aspect, and not inner aspect, the system that is in charge of conveying this information in language. While talking about inner aspectual primitives (such as the case of the "spacetime point"), we are stipulating abstract concepts, which have no reference in the real world.

domain of mathematics and we enter the field of linguistics, we also leave the terrain of coordinates. As far as language is concerned, the idea of "spacetime point" should not be understood as a system of coordinates (x, y, z, t) with possible discrete values, but rather as an abstract bundle of spatiotemporal information, which, perhaps, can be represented cognitively as a point, and linguistically is nothing more than an aspectual primitive, that is to say, an abstract notion that, in turn, encompasses the notions of space and time.

Once we reach this level of abstraction, an important question for linguists would be how such a [spacetime point] primitive is encoded and represented in language, or, better, in predicates. In this regard, there are many answers, depending on the specific theory of aspect we use. A possibility I explored in other works is that the [spacetime point] is encoded in predicates as a formal feature [21,22]. As I showed [23], the treatment of the [spacetime point] as a formal feature (in the sense of Zeijlstra [24–26]) can satisfactorily account for the linguistic phenomena of the diagnostic of eventivity (see section 2), which is exactly what linguists expect from a satisfactory primitive.

4   Conclusions

In this paper, I have proposed a solution for a long-standing problem of linguistics by taking advantage of the concept of "spacetime" and its implications in the understanding of physical events as spacetime points. Once observed that the traditional linguistic account of eventivity as dynamism does not allow a satisfactory study of predicates and their linguistic behaviour, I have suggested relating eventivity to the notion of "spacetime point" and I put forward the idea that the aspectual primitive of events is a spacetime point. This approach would solve the problem under discussion in that it provides an understanding of eventivity that is consistent with the semantic (that is to say, the conceptualization) and the syntactic behaviour of predicates.

This study constitutes a proof of concepts that notions of spacetime from physics can be adapted and applied to linguistics. It is a pilot study that can hopefully benefit research on the spatiotemporal dimension of natural languages and, therefore, the connections between cognition and the physical world. For example, this research could represent a starting point to answer more specific questions regarding the relationship between the linguistic representation of spacetime and reality: what is the nature of the spacetime to which outer aspect refers? Does it coincide with reality or is it an ideal subproduct of our consciousness? Likewise, this proposal can serve as a new input to study other long-debated inner aspectual notions of Events, such as "dynamism" and "telicity".

Acknowledgments

This research has been funded by the Spanish Ministerio de Ciencia e Innovación, project FFI 2014-52015-P.

References

]1] Vendler, Z. Verbs and Times. *Philos. Rev.***66**, 143–160 (1957).

[2] Kenny, A. *Action, Emotion and Will.* (Routledge y Kegan Paul, 1963).

[3] Comrie, B. *Aspect. An Introduction to the Study of Verbal Aspect and Related Problems.* (Cambridge University Press, 1976).

[4] Dowty, D. R. *Word Meaning and Montague Grammar. The Semantics of Verbs and Times in Generative Semantics and in Montague's PTQ.* (Reidel, 1979).

[5] Bennet, M. & Partee, B. H. *Toward the logic of tense and aspect in English.* (Indiana University Linguistics Club, 1978).

[6] Maienborn, C. in *Semantics. An international handbook of natural language meaning* (eds. von Heusinger, K., Maienborn, C. & Portner, P.) 802–829 (De Gruyter Mouton, 2011).

[7] Maienborn, C. On the limits of the Davidsonian approach: The case of copula sentences. *Theor. Linguist.***31**, 275–316 (2005).

[8] Maienborn, C. in *Existence: Semantics and Syntax* (eds. Comorovski, I. & von Heusinger, K.) 107–130 (Springer, 2007).

[9] Luján, M. The Spanish copulas as aspectual indicators. *Lingua***54**, 165–210 (1981).

[10] Bach, E. The Algebra of Events. *Linguist. Philos.***9**, 5–16 (1986).

[11] De Miguel, E. in *Gramática descriptiva de la lengua española* (eds. Bosque, I. & Demonte, V.) 2977–3060 (Espasa Calpe, 1999).

[12] Marín, R. El componente aspectual de la predicación. (Phd thesis, Universitat Autònoma de Barcelona, 2000).

[13] Robinson, M. States, aspect and complex argument structures. in *Edinburgh Linguistic Department Conference '94* (1994).

[14] Carlson, G. N. *Reference to Kinds in English.* (Garland Publishing (1980), 1977).

[15] Milsark, G. L. Existential sentences in English. (MIT, 1974).

[16] Einstein, A. Zur Elektrodynamik bewegter Körper. *Ann. Phys.***17**, 891–

921 (1905).

[17] Einstein, A. Ist die Trägheit eines Körpers von seinem Energieinhalt abhängig? *Ann. Phys.***18**, 639–641 (1905).

[18] Einstein, A. Die Grundlage der allgemeinen Relativitätstheorie. *Ann. Phys.***49**, 769–822 (1916).

[19] Einstein, A. Relativitätsprinzip und die aus demselben gezogenen Folgerungen. *Jahrb. der Radioakt.***4**, 411–462 (1907).

[20] Einstein, A. *Relativity: The Special and General Theory*. (H. Holt and Company, 1916).

[21] Silvagni, F. Ser-I, Estar-S. *Lingue e Linguaggio***14**, 215–232 (2015).

[22] Silvagni, F. A feature that makes the difference: aspectual concord in Romance copular clauses. in *49th Annual Meeting of the Societas Linguistica Europaea* (2016).

[23] Silvagni, F. Entre Estados y Eventos. (Phd thesis in preparation, Universitat Autònoma de Barcelona, 2017).

[24] Zeijlstra, H. There is only one way to agree. *Linguist. Rev.***29**, 491–539 (2012).

[25] Zeijlstra, H. in *The Limits of Syntactic Variation* (ed. Biberauer, T.) 143–174 (John Benjamins, 2008).

[26] Zeijlstra, H. in *Minimalism and Beyond. Radicalizing the interfaces.* (eds. Kosta, P., L. Franks, S., Radeva-Bork, T. & Schürcks, L.) 109–129 (John Benjamins, 2014).

[27] Acedo Moreno, L. La 'kínesis' aristotélica: ¿una actividad abierta? *Scr. Philos. Nat.***1**, 29–56 (2012).

[28] Pustejovsky, J. The Syntax of Event Structure. *Cognition***41**, 47–81 (1991).

[29] Tenny, C. *Aspectual roles and the syntax-semantics interface.* (Kluwer Academic Publishers, 1994).

[30] Verkuyl, H. J. *A Theory of Aspectuality. The Interaction between Temporal and Atemporal Structure.* (Cambridge University Press, 1993).

[31] Davidson, D. in *The Logic of Decision and Action* (ed. Rescher, N.) 81–95 (University of Pittsburgh Press, 1967).

[32] Silvagni, F. Some copular constructions are D-States. in *Chronos XII* (2016).

[33] Minkowski, H. Raum und Zeit. *Phys. Zeitschrift* **10**, 104–111 (1909).

[34] Einstein, A. Bemerkung zur Arbeit von Mirimanoff: Die Grundgleichungen... *Ann. Phys.* **28**, 885–888 (1909).